THE INTERPRETATION
OF ORDINARY LANDSCAPES

Pulaski, New York (Milo Stewart)

THE INTERPRETATION
OF ORDINARY LANDSCAPES

Geographical Essays

D. W. Meinig, Editor
J. B. Jackson
Peirce F. Lewis
David Lowenthal
D. W. Meinig
Marwyn S. Samuels
David E. Sopher
Yi-Fu Tuan

OXFORD UNIVERSITY PRESS
New York **1979** **Oxford**

Copyright © 1979 by Oxford University Press, Inc.

Library of Congress Cataloging in Publication Data
Main entry under title:

The Interpretation of ordinary landscapes.

Seven of the nine essays derive from a special series
of lectures presented at Syracuse University.
Includes index.
1. Landscape assessment—Addresses, essays, lectures.
I. Meinig, Donald William, 1924–   II. Jackson, John
Brinckerhoff, 1909–
GF90.157   301.31   78-23182
ISBN 0-19-502536-9

"Axioms of the Landscape," *Journal of Architectural Education,*
© 1976 by the Association of Collegiate Schools of Architecture,
Inc. Reprinted by permission.

"The Beholding Eye," *Landscape Architecture,* © 1976 by Landscape
Architecture. Reprinted by permission.

Printed in the United States of America

Preface

The set of essays is offered as a coherent introduction to a lively and expanding realm of interest. The nine essays are themselves an exhibit of the vitality of the topic "landscape." Their coherence comes not from any explicit collaboration nor the forcing efforts of an editor, but arises naturally from the fact that these seven very independent writers are well attuned to one another's ideas.

Although we have labeled these essays with our field, we intend no such limitation in their use. To call them "geographical" suggests something of a common perspective, but one which seems to us to be necessarily involved in some degree in almost any approach to "landscape." We have sought to serve a range of interests well beyond the usual bounds of any one academic guild, just as we ourselves have obviously been nourished by a wide variety of writers.

The opening essay is a spirited assertion of some fundamentals by one of America's most successful teachers and writers about landscape. This particular piece is a revised and expanded version of an article published under the title "Axioms of the Landscape" in the special issue on 'Teaching the Landscape" of the *Journal of Architectural Education* (vol. 30, September

1976, pp. 6–9). The second essay, "The Beholding Eye," was prepared originally as the opening lecture for an experimental course on the American Landscape and was first published in *Landscape Architecture* (vol. 66, January 1976, pp. 47–54). Permission to reprint these two articles is gratefully acknowledged.

The remaining essays derive from a special series of public "Landscape Lectures" presented at Syracuse University. I thank my companions for their willingness to join in these explorations and to prepare manuscripts for this volume. A special feature of that program was the showing, for the first time in America so far as we know, of the BBC-2 Television film, "The Making of The English Landscape," based on the work of and featuring as narrator and guide Professor William G. Hoskins. Because the approach and achievements of Professor Hoskins are so central to the theme of this volume I prepared the final essay as a brief assessment of his work and used it also as a means of making some comparisons with the work of Mr. J. B. Jackson, who has been the most widely influential American catalyst of landscape study. I am especially indebted to these two gentlemen for their gracious cooperation. During each of two recent sojourns in Britain I spent a pleasant afternoon with Professor Hoskins in his home in Exeter discussing his work and career. With similar generosity, Brinck Jackson agreed to stop over at my home on one of his transcontinental journeys (this time in a Datsun pickup) and respond to my nagging array of questions about his life's work. I also owe thanks to Peter Jones of BBC-Television, Roy Millward of Leicester University, and Eric Major of Hodder & Stoughton for information about the Hoskins films and books, to Hodder & Stoughton for permission to quote extensively from *The Making of The English Landscape* and *Leicestershire*, and to Blair Boyd for permission to quote many passages from *Landscape*. These good people are of course not to be burdened with any responsibility for the particular interpretations I have made. I thank Nancy Amy and Joyce Berry of Oxford University Press for their skilled help, and Michael Nickerson for assistance in preparing the index. As usual, I have had the ready support in many ways of my departmental Chairman, Professor Robert G. Jensen; both the lecture series and the preparation of the manuscript have been aided by the Cressey-James Fund.

Syracuse, New York   D. W. M.
January 1979

Contents

III American Expressions

THE ORDER OF A LANDSCAPE
Reason and Religion in Newtonian America   J. B. JACKSON

SYMBOLIC LANDSCAPES
Models of American Community   D. W. MEINIG

IV Teachers

READING THE LANDSCAPE
An Appreciation of W. G. Hoskins and J. B. Jackson
D. W. MEINIG

THE CONTRIBUTORS

INDEX

THE INTERPRETATION
OF ORDINARY LANDSCAPES

Introduction

Landscape *is an attractive, important, and ambiguous term.*

It is attractive because it may bring immediately to mind some pleasant prospect: a piece of the countryside, the particular setting of some memorable place; it has an earthy, out-of-doors connotation, it may remind us of environmental or ecological matters; it may suggest special attention to the design and care of our surroundings, or the depiction and interpretation of interesting scenes; to some it may even be regarded as a way of viewing man as well as of admiring nature.

It is an important term because it does carry all these connotations and more, and is thereby involved in major matters of professional interest and of public concern. *Landscape* is a technical term used by artists and earth scientists, architects and planners, geographers and historians. It is also an important dimension of many issues relating to the development, alteration, and management of our cities and towns and countrysides. But beyond all these, it is important because it is a common word which is increasingly used to encompass an ensemble of ordinary features which constitute an extraordinarily rich exhibit of the course and character of any society. As Peirce Lewis spiritedly argues, if we want to understand ourselves, we would do well to take a searching look at our landscapes.

Because *landscape* is used by so many different people for such a variety of purposes, it is inevitably an ambiguous term. There are problems of translation between fields and often uncertainties of exact meaning even within any one. When J. B. Jackson, our most catholic and discerning spokesman, who bound a good portion of his life directly to the very word, confesses (as he does in this book) that after twenty-five years he still finds the concept of landscape elusive, we are fairly warned not to aspire to a clean and clear definition, and not to be surprised at some variation in usage among the seven authors of this volume. Nevertheless, the reader deserves some indication of the general sense of the term which informs these essays. In the hope of clarification we shall begin by differentiating *landscape* from some closely related concepts; to say first of all what it is not. This is done not to establish rigid restrictions to its use, but to distill something closer to an essence which can be put to good and varied service.

Landscape is related to, but not identical with, *nature*. Nature is a part of every landscape, but is no more than a part of any landscape which has felt the impact of man. In this view landscape is always inclusive of man and nature, rather than a way of distinguishing, or at least emphasizing, nature, as is still not uncommon in some fields, such as art and earth science. Indeed, the idea of landscape runs counter to recognition of any simple binary relationship between man and nature. Rather, it begins with a naive acceptance of the intricate intimate intermingling of physical, biological, and cultural features which any glance around us displays. Landscape is, first of all, the unity we see, the impressions of our senses rather than the logic of the sciences.

Thus every landscape is a scene, but landscape is not identical with *scenery*. The very idea of scenery is limited, a conscious selection of certain prospects, locales, or kinds of country as having some attractive aesthetic qualities. Scenery has connotations of a set piece, a defined perspective, a focus upon certain features, a discrimination based upon some generally received idea of beauty or interest; whereas landscape is ubiquitous and more inclusive, something to be observed but not necessarily admired. Interest in landscape may involve aesthetics but it is not defined by it. As with landscape art, the study of landscape is necessarily reflective in some degree of philosophies and taste and subject to shifts in styles and emphasis, but the landscape is ever with us and we are ever involved in its creation.

Landscape is all around us. It is related to, but not identical with, *environment*, as several of the authors make explicit. Environment is an inherent property of every living thing, it is that which surrounds and sustains; we are always environed, always enveloped by an outer world. Landscape is less inclusive, more detached, not so directly part of our organic being. Landscape is defined by our vision and interpreted by our minds. It is a panorama which continuously changes as we move along any route. Strictly speaking, we are never in it, it lies before our eyes and it becomes real only as we become conscious of it. As Tuan says, "We can think, therefore we are able to see an entity called landscape." Environment sustains us as creatures; landscape displays us as cultures.

As discerned sectors of our environments, landscapes are related to, yet not identical with, *places*. Place commonly refers to a definite area, a fixed location; events "take place" and we can be *in* a place. But place, too, has its ambiguities. There is, most basic of all, the difference between general recognition of certain areas as places, and a personal sense of place. The one is a public concept, the other private; we all live intimately with both. The first kind of place depends upon some public agreement as to name, location, and character; some legibility, some identity commonly understood. Our personal sense of place depends upon our own experiences and sensibilities. It is unique to each of us in its content and in the way it relates to general social definitions of places. Thus each of us creates and accumulates places out of living whenever we pierce the infinite blur of the world and fix a piece of our environment as something distinct and memorable. Such memories of place almost certainly depend in some degree upon landscape, upon the external visible character of localities. Yet the two are not the same. As David Sopher suggests, for some people the sense of home as a place may be grounded much more upon human relationships than upon the memory of landscape. In this way place is experiential to a degree landscape is not, although the way we see landscape does depend upon experience and purpose. Still landscape tends to be something more external and objective than our personal sense of place; and something less individual, less discrete, than the usual named place; it is a continuous surface rather than a point, focus, locality, or defined area.

Landscape is a portion of the earth's surface, related to, but not identical with, *region, area,* or *geography*. There are complexities and ambigu-

We regard all landscapes as symbolic, as expressions of cultural values, social behavior, and individual actions worked upon particular localities over a span of time. (P.J. Hugill)

ities here which have been important and at times vexing in the field of geography. Such problems grow out of its long and rich heritage, they reflect shifts in Western intellectual history, differences in emphasis among nations, difficulties of translation between languages, and the differential impact of allied fields. These matters are well covered in standard references[1] and are far too complicated to review here. We may note that there was a period between the World Wars when many American geographers tried to define their discipline in terms of landscape. That proved to be a stimulating but not, in the longer run, satisfying concept and it is obvious from current professional literature that while landscape remains an important focus of geographical interest the field itself could not possibly be comfortably encompassed within the bounds of common concepts of landscape. And this is so not just because of a recent emphasis upon geography as a more theoretical spatial science, but because of the special analytical perspective which has always been characteristic of the field. That oldest

In its focus upon the vernacular, cultural landscape study is a companion of that form of social history which seeks to understand the lives of ordinary people. (P.J. Hugill)

badge and basic tool of geography, the map, is a symbolic abstraction of spatial relationships and is applied by geographers to the study of many phenomena which are not directly part of the visible landscape. On the other hand, maps may be useful in the study of landscape but they cannot be sufficient, for the landscape must be visualized and if not directly by our own eyes then by means of the best substitutes. The photograph or drawing, the depiction of the surficial totality of a scene, provides a more revealing illustration than a map. Nevertheless the relationship between landscape and geography remains intimate even if noncommensurate. As David Lowenthal has noted, "beyond that of any other discipline . . . the subject matter of geography approximates the world of general discourse; the palpable present, the everyday life of man on earth, is seldom far from our professional concerns."[2] On the basis of logic, tradition, and product, we may fairly claim that geographers have a special vocation for landscape study.

Our concern in these essays for that everyday life of man on earth is indicated in our title. We specify *ordinary landscapes* to indicate our primary interest in that continuous surface which we can see all around us. We cannot, of course, study everything, but we can try to see those elements we do study in context, as being parts of an ensemble which is under continuous creation and alteration as much or more from the unconscious processes of daily living as from calculated landscape design. Insofar as we focus on particular landscapes, we are dealing primarily with vernacular culture. In this sense, landscape study is a companion of that form of social history which seeks to understand the routine lives of ordinary people. And indeed, the relationship with social history is even closer for although we begin with the "palpable present," with that which we can see, *interpretation* will demand more than can be seen in a mere glance and a concern for more than the palpable objects themselves. For the meaning of the ordinary is rarely obvious. We regard all landscapes as symbolic, as expressions of cultural values, social behavior, and individual actions worked upon particular localities over a span of time. Every landscape is an accumulation, and its study may be undertaken as formal history, methodically defining the making of the landscape from the past to the present, as in the great work of W. G. Hoskins and his associates. And every landscape is a code, and its study may be undertaken as a deciphering of meaning, of the cultural and social significance of ordinary but diagnostic features, as shown in numerous revealing essays by J. B. Jackson.

It is not, however, the intent of the essays in this book to prescribe any exact form of study but to explore possibilities and to invite others to do the same. It is an immense realm which needs many kinds of explorers. Any landscape is so dense with evidence and so complex and cryptic that we can never be assured that we have read it all or read it aright. The landscape lies all around us, ever accessible and inexhaustible. Anyone can look, but we all need help to see that it is at once a panorama, a composition, a palimpsest, a microcosm; that in every prospect there can be more and more that meets the eye.

Notes

1. Preston E. James, *All Possible Worlds, A History of Geographical Ideas* (New York: Odyssey Press, 1972), pp. 229–32, 399–402; Marvin W. Mikesell, "Landscape,"

International Encyclopedia of the Social Sciences, vol. 8, (New York: Crowell-Collier and Macmillan, 1968), pp. 575–80.

2. David Lowenthal, "Geography, Experience, and Imagination: Towards a Geographical Epistemology," *Annals*, Association of American Geographers 51 (1961): 241.

I

Fundamentals

Axioms for Reading the Landscape

Some Guides to the American Scene

Peirce F. Lewis

About the axioms and about cultural landscape

For most Americans, ordinary man-made landscape is something to be looked at, but seldom thought about. I am not talking here about "natural landscape," but about the landscape made by humans—what geographers call *cultural landscape.* Sometimes Americans may notice cultural landscape because they think it is pretty, or perhaps ugly; mostly they ignore the common vernacular scene. For most Americans, cultural landscape just *is.*

Usage of the word tells a good deal. As a common verb, to "landscape" means to "prettify." If a suburban lot is advertised as "landscaped," it is generally understood that somebody has fussed with the shrubbery on a small bit of ground, perhaps planted a few trees, and has manicured the bushes—more or less artfully. It rarely occurs to most Americans to think of landscape as including everything from city skylines to farmers' silos, from golf courses to garbage dumps, from ski slopes to manure piles, from millionaires' mansions to the tract houses of Levittown, from famous historical landmarks to flashing electric signs that boast the creation of the 20 billionth hamburger, from mossy cemeteries to sleazy shops that sell pornography next door to big city bus stations—in fact, whole countrysides, and whole cities, whether ugly or beautiful makes no difference. Although the word is seldom so used, it is proper and important to think of cultural landscape as nearly everything that we can see when we go outdoors. Such common workaday landscape has very little to do with the skilled work of landscape architects, but it has a great deal to say about the United States as a country and Americans as people.[1]

At first, that idea sounds odd. The noun "landscape" evokes images of snow-capped mountains and waves beating on a rock-bound coast. But the fact remains that nearly every square millimeter of the United States has been altered by humankind somehow, at some time. "Natural landscapes" are as rare as unclimbed mountains, and for similar reasons. Mallory expressed a very American sentiment when he said he wanted to climb Everest because it was there. Americans tinker with landscape as if pursued by some inner demon, and they have been doing so ever since their ancestors landed at Jamestown and Plymouth and began chopping down trees. They continue today, and the sound of the power lawn-mower is heard throughout the land.

All of this is obvious, but the implications are less obvious, though very simple, and very important to our understanding of the United States. The basic principle is this: *that all human landscape has cultural meaning*, no matter how ordinary that landscape may be. It follows, as Mae Thielgaard Watts has remarked, that we can "read the landscape" as we might read a book.[2] Our human landscape is our unwitting autobiography, reflecting our tastes, our values, our aspirations, and even our fears, in tangible, visible form. We rarely think of landscape that way, and so the cultural record we have "written" in the landscape is liable to be more truthful than most autobiographies because we are less self-conscious about how we describe ourselves. Grady Clay has said it well: "There are no secrets in the landscape."[3] All our cultural warts and blemishes are there, and our glories too; but above all, our ordinary day-to-day qualities are exhibited for anybody who wants to find them and knows how to look for them.

To be sure, reading landscapes is not as easy as reading books, and for two reasons. First, ordinary landscape seems messy and disorganized, like a book with pages missing, torn, and smudged; a book whose copy has been edited and re-edited by people with illegible handwriting. Like books, landscapes *can* be read, but unlike books, they were not *meant* to be read.

Our human landscape is our unwitting autobiography, and all our cultural warts and blemishes, our ordinary day-to-day qualities, are there for anybody who knows how to look for them.

In the second place, most Americans are unaccustomed to reading landscape. It has never occurred to them that it *can* be done, that there is reason to do so, much less that there is pleasure to be gained from it. That is one reason why so many Americans prefer driving on freeways with their bland highway-department roadsides, to driving on old-fashioned roads with their curves and crossroads and billboards and towns and irresponsible pedestrians and cyclists and straying livestock and roadside houses that spew forth children chasing balls—in short, all the things that make driving backroads interesting and hazardous. Very few academic disciplines teach their students how to read landscapes, or encourage them to try. Traditional geomorphology and traditional plant ecology (and one must, alas, stress "traditional" here) were two happy exceptions: these were disciplines which insisted that their practitioners use their eyes and *think* about what they saw, and it is no accident that some of America's most accomplished landscape-readers, such as J. Hoover Mackin, Pierre Dansereau, and Mae Thielgaard Watts, derive from those fields.[4] A few cultural geographers are also noteworthy. In America, much of the inspiration derived from Carl Sauer, who built the remarkable and influential "Berkeley School" in geography at the University of California, and whose students number some of the most accomplished landscape-readers in American professional geography.[5] Fred Kniffen[6] comes immediately to mind, as do Wilbur Zelinsky,[7] David Lowenthal[8] and James Parsons.[9] One thinks, too, of the Minnesota geographers, John Fraser Hart,[10] Cotton Mather, and Harry Swain.[11] But the list of geographers is not long. More often, you run across accomplished landscape-readers in unexpected places. J. B. Jackson, founder and long-time editor of *Landscape* magazine, diffidently disclaims association with any particular discipline; his work is dazzling and his influence (inside and outside academe) has been profound.[12] Henry Glassie, the folklorist and student of Kniffen, is another.[13] George Stewart, one of the best, hung his academic hat in the English Department at Berkeley, and we are all the richer for it.[14] Some journalists are among the most perceptive, perhaps because they spend their lives looking and writing about what they see, no matter how trivial the subject may seem. Tom Wolfe argues that police-reporters of the old school are among our most potent social analysts—the old fellows in battered felt hats who walked through life with the wide eyes of cynical children, noting everything in a curiously innocent way, and writing about what they saw.[15] But it remains a sad fact that most academics in those fields where we might expect to find expert landscape-reading are egregiously inept. To be sure, there are glorious exceptions—people like Alan Gowans, Reyner Banham, and Grady Clay—but they remain exceptions nonetheless.

So unless one is lucky enough to have studied with a plant ecologist like Dansereau, a geomorphologist like Mackin, a folklorist like Glassie, or simply a Renaissance man like Jackson, one is likely to need guidance. To "read landscape," to make cultural sense of the ordinary things that constitute the workaday world of things we see, most of us need help.

I took a long time learning that fact. Years ago, when I started teaching about cultural landscapes of the United States, I was puzzled and annoyed that students seemed so obtuse. They seemed blind to all that marvellous material around them, and even worse, some of them seemed insulted when they were told to go outdoors and use their eyes and think about what they saw. Gradually, I realized that the students were not obtuse; *I* was. The students were simply aping the great majority of their faculty mentors, who by their inattention to ordinary landscape were teaching the students very effectively that landscape didn't matter: that serious students did not deal with trivial questions about ordinary everyday things, such as what Howard Johnsons's cupolas were meant to symbolize, or why people put pink plastic flamingos in their front yards.

What we needed, I concluded, were some guides to help us read the landscape, just as the rules of grammer sometimes help guide us through some particularly convoluted bit of syntax. Little by little, I began to write down some of the rules that I discovered over the years of looking and learning and teaching about American landscapes and which I found helped me understand what I saw. I call these rules "axioms," because they now seem basic and self-evident, as any proper axiom must be. I may be wrong in using the word "axiom": what seems self-evident now was not obvious to me a few years ago. But call them what you will: They are nevertheless essential ideas that underlie the reading of America's cultural landscape.

The Axioms

1. THE AXIOM OF LANDSCAPE AS CLUE TO CULTURE *The man-made landscape—the ordinary run-of-the-mill things that humans have created and put upon the earth—provides strong evidence of the kind of people we are, and were, and are in process of becoming.* In other words, the culture of any nation is unintentionally reflected in its ordinary vernacular landscape.

THE COROLLARY OF CULTURAL CHANGE Our human landscape—our houses, roads, cities, farms, and so on—represents an enormous investment of money, time, and emotions. People will not change that landscape unless they are under very heavy pressure to do so. We must conclude that if there is really major change in the *look* of the cultural landscape, then there is very likely a major change occurring in our national culture at the same time.

THE REGIONAL COROLLARY If one part of the country (or even one part of a city) *looks* substantially different from some other part of the country (or city), then the chances are very good that the cultures of the two places are different also. Thus, much of the South looks different from the rest of the country, not only because the climate and vegetation are different, but also because some important parts of Southern culture really *are* different from the rest of the country, although not necessarily in the way that some propagandists would like us to think. So also, black ghettos in Northern cities look different from adjacent white slums, because the culture of such ghettos remains distinctive.

THE COROLLARY OF CONVERGENCE To the degree that the look of two areas comes to be more and more alike, one may surmise that the cultures are also converging. Thus many small Southern towns look quite different from their Northern counterparts, while Atlanta looks more and more like the "standard" Northern city, and even something like Phoenix, which is perhaps American's most super-American city. One may properly conclude that the cultural rift between North and South is growing narrower, but the process of reunion is taking place faster in urban places than in rural ones, and fastest of all in the suburbs.

To take another example: black suburbs of Northern cities look increasingly like white suburbs, and the shacks of rural Southern blacks are simultaneously being replaced by replicas of the "ranchettes" of exurban Northern whites. It may be legitimate to speculate that such convergence of landscapes represents some real convergence of cultures and perhaps some lessening of racial tensions.

THE COROLLARY OF DIFFUSION The look of a landscape often is changed by imitation. That is, people in one place see what is happening elsewhere, like it, and imitate it if possible. The timing and location of such imitative changes are governed by various forms of geographic and social diffusion, which are surprisingly predictable, and which tell us a good deal about the way that cultural ideas spread and change. For example, Greek Revival architecture spread from Virginia into upstate New York in the early nineteenth century and from there, in debased form, to other parts of the country. Both the spread and the debasement took nearly a century to complete.[17] Now, in the 1970s, California landscape tastes are widely and wildly imitated in most parts of the country. The delay between California invention and Eastern imitation is extremely small—sometimes almost instantaneous.

Much of our ordinary landscape reflects vagaries of fashion and taste which may be rapidly and widely diffused. Why does a pseudo-Spanish mission-style building show up on an automobile row in Syracuse, New York? (P.J. Hugill)

THE COROLLARY OF TASTE   Different cultures possess different *tastes* in cultural landscape; to understand the roots of taste is to understand much of the culture itself.

While most people admit they have "taste" in landscape, and in fact would insist they do, they often claim that their tastes are based on "practical" grounds. That is ludicrously untrue in most instances. A huge amount of our day-to-day behavior and the landscapes created by that behavior is dictated by the vagaries of "fashion" or "taste" or "fad." And when we speak of "taste," we are talking about *culture*—not about practicality.[18]

At first glance, some fads seem trivial, like hula-hooping or skateboarding: apparent eccentricities that sweep the country and then are gone. But what guides those fads? Are they really so differ-

ent from the deep-seated cultural biases that anthropologists and
cultural geographers take so seriously: dietary "laws" that encourage
us to eat the meat of steers and chickens, and produce nausea at the
thought of eating rats and snakes? Why do we build domes and spires
on public buildings, but rarely on our houses? Why did lightning rods
suddenly appear on the American scene, and then disappear except
as antiques? Why do we plant our front yards to grass, water it to
make it grow, mow it to keep it from growing too much, and impose
fines on those who fail to mow often enough? (Why not let the dande-
lions grow, or pour concrete instead? Occasionally people do just
that, and are ostracized by their neighbors.) At best, the answers to
these questions are subtle, fascinating, and often very hard to get. At
worst, we simply have no answers at all. But we know enough about
taste to know that it is a powerful cultural force (avoiding rat-eating,
for example), and those tastes do not come about by accident. Indeed,
to trace the paths of taste through historic time and across geographical space tells us a good deal about the nature of American
cultures: what it is, and how it got to be the way it is.

Thus, if we ask why America's human landscapes look the way
they do, it may seem that we are asking simple-minded questions. In
fact, we are also asking: Why do Americans possess certain tastes and
not others? We are asking where those tastes came from and why
they take hold in certain times and disappear at others.

2. THE AXIOM OF CULTURAL UNITY AND LANDSCAPE EQUALITY *Nearly all
items in human landscapes reflect culture in some way. There are almost no
exceptions. Furthermore, most items in the human landscape are no more and
no less important than other items—in terms of their role as clues to culture.*
Thus, the MacDonald's hamburger stand is just as important a cultural symbol (or clue) as the Empire State Building, and the change in
design of MacDonald's buildings may signal an important change in cultural attitudes, just as the rash of Seagram's "shoebox skyscrapers" around
exurban freeway interchanges heralds the arrival of a new kind of American
city—and a new variant of American culture. So also the painted cement
jockeyboy on the front lawn in lower middle-class suburbia is just as important as a symbol as the Brooklyn Bridge; the Coney Island roller rink is as
important as the Washington Monument—no more, no less.

This axiom parallels an equally basic proposition: that culture is
whole—a unity—like an iceberg with many tips protruding above the sur-

face of the water. Each tip looks like a different iceberg, but each is in fact part of the same object. The moral is plain: no matter how ordinary it may seem, there's no such thing as a culturally uninteresting landscape.

But note these caveats:

a. If an item is really unique (like the only elephant-shaped hotel south of the 40th parallel, located in Margate, New Jersey), it may not seem to mean much, except that its creator was rich and crazy.

b. *However,* one should not be too hasty in judging something "unique." That elephant-shaped hotel has many close relatives: giant artichokes in Castroville, California; billboards that blow smoke rings in Times Square. In some circles such things are called "camp" or "pop" or "kitsch," and it is fashionable to snicker at them. But ridicule or deprecation cannot dismiss the persistent, nagging and fascinating question: what do these ordinary things tell us about American culture?[19]

c. The fact that all items are equally important emphatically does *not* mean that they are equally easy to study and understand (cf. Axiom 7). Sometimes the commonest things are the hardest to study; which leads us to . . .

3. THE AXIOM OF COMMON THINGS *Common landscapes—however important they may be—are by their nature hard to study by conventional academic means.* The reason is negligence, combined with snobbery. One has no trouble finding excellent books about famous buildings like Monticello or famous symbolic structures like the Brooklyn Bridge.[20] Curious antique objects get a lot of attention too: "olde" spinning wheels and "Olde" Williamsburg. But it is hard to find intelligent writing which is neither polemical nor self-consciously cute on such subjects as mobile homes, motels, gas stations, shopping centers, billboards, suburban tract housing design, the look of fundamentalist churches, watertowers, city dumps, garages and carports. Yet such things are found nearly everywhere Americans have set foot, and they obviously reflect the way ordinary Americans think and behave most of the time. It is impossible to avoid the conclusion that we have perversely overlooked a huge body of evidence which—if approached carefully and studied without aesthetic or moral prejudice—can tell us a great deal about what kinds of people Americans are, were, and may become.[21]

THE COROLLARY OF NONACADEMIC LITERATURE Happily, not all American writers, nor foreign visitors, are as snooty as American scholars. Even though there is little written about motels and fast-foot eateries in the "standard" scholarly literature, the country is awash with fascinating and useful material about these common items. One merely has to look in the right place. Some of the "right places" include:

a. Writings of the "new journalists," like Tom Wolfe, who reflect with devastating accuracy on such things as the landscape of drag racing, Las Vegas billboards, the architecture of surfing (including surfers' arcane haircuts), and above all, the cultural contexts from which such landscapes spring.[22]

b. Trade journals, written for people who make money from vernacular landscapes. If, for example, you want to know why your local franchised hamburger joint looks the way it does, try browsing through the pages of the journal *Fast Food*. The magazine is intended for and read by a very select audience of restaurateurs and investors, and it contains advice to help its readers get rich as fast as possible. Large sections of the journal deal with such matters as the workings of space-age ovens that fry quarter-pound hamburgers instantly, but nestled among the technical esoteria, the student of vernacular landscape can find a treasury of cultural information. There is remarkably candid advice on restaurant design that has been road tested to catch the traveler's eye: outdoor signs and landscaping formulas that are based on cool, even chilly appraisals of American popular taste, a matter that lies at the very roots of culture. Then too, trade journals often contain a page or two of "political news" (often called something like "Washington Hotline") which report on political and legal matters of importance to the industry in question. Thus, a trade journal for highway engineers and road contractors, *Rural and Urban Roads* (previously called *Better Roads*), reports on congressional hearings which may result in new laws which will in turn determine what our road systems and roadsides will look like for years to come. The journal often urges subscribers to support legislation that has not even been written yet, and it suggests ways to promote that legislation by influencing politicians, inserting "plugs" in the popular press, and generally engaging in what is politely termed "PR".

For the would-be landscape-reader—one who is neither restaurateur nor highway engineer—browsing through those trade journals can be disconcerting. It is rather like stumbling accidentally across a highly classified document that outlines detailed plans for a military invasion across home territory. Not only does one learn what our future landscapes may look like, one learns in advance about some of the methods by which they may be created. It is not often that scholars in any field have such a chance to look into the future.

Trade journals, especially the old, established ones, can usually be trusted for their cultural judgments. (If their appraisals are wrong too often, they simply do not stay in business.) There are hundreds, indeed thousands, of such journals, and they are not hard to find.[23]

c. Advertisements for commercial products. One need not speculate very long to identify the strain in the American psyche that the obviously successful ad-makers for Marlboro cigarettes are trying to touch. By whatever name it is called, Marlboro's wild-west country has a very real place in America's collective landscape tastes, and those tastes emerge in some very real places: fire departments that look like pueblos in the suburbs of Buffalo, New York; "Western stores" in eastern Louisiana; and "desert lawns" (replete with sand, cactus, bleached wood, and longhorn skulls) spread from arid Tucson to the foggy shores of San Francisco Bay.[24] Old advertisements are equally valuable, for they speak volumes about past technology, past taste, and past cultures. In the same way, old illustrations, picture postcards, or photgraphs may serve similar purposes.

d. Promotional travel literature, often in the form of slick brochures that tell you not very subtly what you are supposed to see when you visit certain places. Recent changes in the landscape of the Pocono Plateau, for example, are much easier to understand after one has seen the marginal eroticism in brochures that beckon newlyweds to any of several Pocono "honeymoon retreats." Indeed, travel literature can *act* as an agent of landscape change. Much of New Orleans's French Quarter, for example, has been "upgraded" and sanitized so that it would accord with tourists' expectations. Those expectations, of course, largely derive from advertising which has been directed at the tourist himself. The advertisement thus becomes a self-fulfilling prophesy.

e. The rare book by a perceptive person who has looked intently at a landscape and discovered what it means. If one really wants to understand what Americans are doing and thinking and aspiring to, sample the glories of George Stewart's *U.S. 40: Cross-section of the United States of America* (1953)[25] or Grady Clay's superb *Close-up: How to Read the American City* (1973)[26] Almost anything by J. B. Jackson will do the job nicely, although "The Stranger's Path" is especially perceptive.[27]

4. THE HISTORIC AXIOM *In trying to unravel the meaning of contemporary landscapes and what they have to "say" about us as Americans, history matters.* That is, we do what we do, and make what we make because our doings and our makings are inherited from the past. (We are a good deal

A large part of the common American landscape was built by people in the past, whose tastes, habits, technology, wealth, and ambitions were different from ours. Southern Illinois. (P.F. Lewis)

more conservative than many of us would like to admit.) Furthermore, a large part of the common American landscape was built by people *in* the past, whose tastes, habits, technology, wealth, and ambitions were different than ours today. Thus, while we live among obsolete artifacts of past times—"old-fashioned houses" and "obsolete cities" and "inefficient transportation" or "bad plumbing"—those objects were not seen to be "inefficient" or silly by the people who made them, or caused them to be made. To understand those objects, we must try to understand the people who built them—our cultural ancestors—in *their* cultural context, not ours.

THE COROLLARY OF HISTORIC LUMPINESS Most major cultural change does not occur gradually, but instead in great sudden historic leaps, commonly provoked by such great events as wars, depressions, and major inventions. After these leaps, landscape is likely to look very different than it did before. Inevitably, however, a lot of "pre-leap" landscape will be left lying around, even though its reason for being has disappeared. Thus, the Southern landscape is littered with sharecroppers' houses, even though the institution of sharecropping has nearly disappeared—a victim of the boll weevil and a concatenation of other forces that combined to destroy the old Cotton Belt of the early 1900s, and provoked a migration of black farmers northward, eventually to change the entire urban landscape of industrial America. Most small towns in America—at least of the Norman Rockwell ilk—are like the Cotton Belt: obsolete relics of a different age. There are no more being built today, and, unless things in America change radically, there never will be.

THE MECHANICAL (OR TECHNOLOGICAL) COROLLARY To understand the cultural significance of a landscape or an element of the landscape, it is helpful (and often essential) to know in *particular* about the *mechanics* of technology and communications that made the element possible.

    For example, we can speculate endlessly (and often pointlessly) about the "symbolism" of, say, the American front lawn, made of mown green grass: perhaps it is a status symbol reflecting a borrowing from England, and thus a subliminal reflection of our admiration for things English. But much of that "symbolic speculation" is likely to be hot air unless we really know *how* a lawn works in a very mechanical way. The fact that most of us have direct experience with

lawns, planting and mowing and fertilizing and irrigating and curs-
ing them, obscures two important facts: 1. We do many mechanical
things to establish and maintain a lawn that we take for granted
(such as getting the lawnmower serviced), but which are nonetheless
essential and that would baffle people from lawless societies. Except
for companies like the Scott Grass Seed Co., nobody bothers to write
about such behavior, commonplace as it is. 2. We need to know who
invented the machinery to make the lawn possible: who took that
invention and engineered the machinery so that it came within the
financial reach of Everyman (invention and engineering are emphatically not the same thing); who adopted the machinery; how the idea
spread; and above all, *when* all this happened and in what order; and
where these events took place and how they spread, often in direct
defiance of environmental good sense. (Why are there green lawns all
over Sun City, Arizona, for example? And why, only recently, the
sudden efflorescence—if one may call it that—of those "desert lawns"
throughout the West?)

   All that, of course, is a big order for something so commonplace as the American lawn. Yet pause and consider what we are
really discussing. Every step of the way we are investigating the
evolution of American culture: where things started, when, and how.
The key work is *how*, for unless one knows about the technology
behind the landscape element we are concerned with, the fact remains that we really know very little about it. Speculation about
symbolisms will remain unprofitable.[28]

5. THE GEOGRAPHIC (OR ECOLOGIC) AXIOM *Elements of a cultural landscape make little cultural sense if they are studied outside their geographic (i.e.,
locational) context.*

   To a large degree, cultures dictate that certain activities should occur
in certain places, and only in those places. Thus, all modern American cities
are segregated: streetwalkers are not found throughout the city, nor are
green lawns, trees, high buildings, or black people. This axiom is so obvious
that it should not have to be mentioned, except that so many scholars and
"practical" people persistently flout it. Architectural historians publish
books full of handsome photographs of "important buildings," artfully composed so that the viewer will not see the "less important" building next
door, much less the telephone wires overhead or the gas station across the
street. The "important building" is disembodied, as if on an architect's

easel in a windowless studio somewhere. So also, planners make grand schemes to improve sections of existing cities, plans drawn on large blank sheets of paper, with adjacent areas shown in vague shades of gray or not shown at all, as if the planning district existed *in vacuo*. The planners are perplexed when residents of those gray areas rise up in anger, and perplexity turns to frustration when city councils send the elegant plans back to rest ignominiously in a file drawer, full of similar material, rejected from the past. Again and again, historic preservationists throw up white picket fences around "historic buildings," while adjacent neighborhoods go to ruin. Inside is "history"; outside, it isn't history.[29] (Then we wonder why the general public equates historic preservation with Disneyland!) To study a building as if it were on an artist's easel, detached from its surrounding, is to remove some of the most important evidence explaining why the building looks the way it does, and what its appearance has to tell us about the culture in which it was built.

It is easy to understand why buildings (for example) are isolated for study outside their geographic surroundings. It is what scientists call a "simplifying assumption," and it makes things easier for the student. So, the epidemiologist studies a deadly microbe in an antiseptic pan of agar so that he can see how the bug behaves in isolation. Thus, he meets the bug. But he knows enough to realize that the microbe is important only *in context*, because it causes the disease in a larger body; in this instance, the environment of the human body. So it is with houses and barns and lawns and sidewalks and any other "item" in the landscape: to make sense of them, one must observe them in context.

6. THE AXIOM OF ENVIRONMENTAL CONTROL *Most cultural landscapes are intimately related to physical environment. Thus, the reading of cultural landscape also presupposes some basic knowledge of physical landscape.*

We often boast that we have "conquered geography," meaning that contemporary technology is so powerful that we can build anything, wherever we like, and effectively ignore climate, landforms, soils, and the like. To be sure, we grow tomatoes in greenhouses all winter long, and Pennsylvanians flee to Florida when their native winters grow excessively obnoxious. We send men to the moon, and we build superhighways almost anywhere we want.

But "conquering geography" is often very expensive business. Compare the price of tomatoes in January with the price in August (and compare the quality, too!), or contrast the cost per mile of a crosstown expressway in New York with one across North Dakota prairies. In earlier simpler times, with less money, less sophisticated tools, and less information, "conquering geography" was even more expensive, and people avoided such extravagance whenever they could. Thus, the South differed culturally from the North largely because it differed physically. Southern cities stopped looking Southern about the time that cheap air conditioning made it possible to ignore the debilitating heat of a super-tropical summer, which lasted sometimes for five months, a season in which nobody who could help it did any work between noon and 7 P.M. The "Southern way of life" was renamed "the Atlanta spirit" and began to take on Yankee ways, largely because of air conditioning. Then the Arabs tripled the price of oil, and suddenly air conditioning became "uneconomical." Sitting on verandahs came back into style, and glass-lined offices in high-rise skyscrapers with windows that wouldn't open were seen as something less than Paradise on an August afternoon. Environment continues to matter after all.

7. THE AXIOM OF LANDSCAPE OBSCURITY *Most objects in the landscape— although they convey all kinds of "messages"—do not convey those messages in any obvious way.* The landscape does not speak to us very clearly. At a very minimum, one must know what kinds of questions to ask:

As for asking questions, one can quickly get into the habit of asking them simply by doing so. What does it look like? How does it work? Who designed it? Why? When? What does it tell us about the way our society works? (It is remarkable how many intelligent perceptive people have never asked questions of the landscape, simply because nobody ever suggested they do it.)

As for the answers, and judging their validity, that is a trickier matter. Many historians, geographers, and others will ask the obvious question: "If you want to interpret American culture, why not simply read books about it? Why use landscape as evidence, especially when you have already admitted that the interpretation of cultural landscape is such a slippery uncertain enterprise?"

There are two answers to this not-so-simple question:

1. Many of the books are not yet written. For example, I know of no satisfactory book about the landscape of recreation in America despite the fact that we spend billions of dollars on recreation every year, that in many places it is the chief source of revenue, and that most Americans spend huge chunks of time either having fun or thinking about it. To be sure, there is no dearth of books about "recreation planning"—solemn tomes about parks

and playgrounds—but if one wants to know about how American tastes have changed through time, one finds the bookshelves nearly empty.[30] Visible evidence is nearly all we have; however, the visible evidence is plentiful: everything from abandoned amusement parks to Little League baseball fields to the little signs stuck on telephone poles in Minnesota and upstate New York that admit that snowmobiles have priority in much of the Northland in winter and thereby admit existence of a subculture that did not exist a decade ago.

2. Many books about certain important subjects (e.g., why American houses look the way they do) disagree with each other, and not in minor ways, either. One must conclude that somebody is not telling the whole truth. The most immediate way to resolve such disagreement is to go back to the real thing (in this case, the house itself). The chances are excellent that part, if not all, of the difficulty can be cleared up by visible evidence (and we will begin to have a growing suspicion that many authors have never looked closely at what they write about).

One can, of course, claim too much for the virtues of landscape-reading. It is not a panacea, not the master key to an understanding of culture. Indeed, it may be no more than a diverting game, because it is pleasant to go outdoors and let your eyes roam idly across some nice bit of scenery and tell yourself that you are engaged in research. (Landscape-reading will not put libraries out of business.)

One can, however, quite literally teach oneself how to see,[31] and that is something that most Americans have not done and should do.[32] To be sure, neither looking by itself, nor reading by itself is likely to give us very satisfactory answers to the basic cultural questions that landscape poses. But the alternation of looking, and reading, and thinking, and then looking and reading again, can yield remarkable results, if only to raise questions we had not asked before. Indeed, that alternation may also teach us more than we had ever dreamed: that there is order in the landscape where we had seen only bedlam before. That may not be the road to salvation, but it may be the road to sanity.

Notes

1. I am talking about Americans in this essay because I am an American and know my countrymen better than the citizens of any other nation. (I may also be talking about Canadians or Australians too, but they can decide that better than I.) But to talk about America is to talk with some meaning for a larger world.

American traits and American landscapes are imitated by other nations, not so much because the traits are American, but because sophisticated technology is at popular disposal more commonly in America than in most parts of the world to produce some very comfortable living. Many dislike the usage, of course, but in some places, the word "American" serves as shorthand for "modern-efficient-comfortable."

2. See Mae Thielgaard Watts, *Reading the Landscape: An Adventure in Ecology* (New York: Macmillan, 1957) Republished (1975) as *Reading the Landscape of America.* See esp. the last chapter, "The Stylish House; or Fashions as an Ecological Factor," where she traces the career of a New England cottage from its construction in the early 19th century to the present, showing how changes in the house, its landscaping, and its residents (all named and numbered, even unto the dog, Fido) kept up with fashions from generation to generation. For a succinct, beautifully illustrated history of domestic middle-class American tastes, I know of nothing like it. Both writing and drawing are inspired. A similar delight awaits in her reading *Reading the Landscape of Europe*, published in Britain as *The Countryside Around You* (London: Cassel, 1973), which opens with a marvelous, illustrated prologue, "Reading the Rooflines of Europe."

3. Grady Clay, *Close-up: How to Read the American City* (New York: Praeger, 1973).

4. J. Hoover Mackin, late Professor of Geomorphology at the University of Washington and Texas, is internationally known for his epochal essay, "Concept of the Graded River," *Bulletin*, Geol. Soc. of America (1948):463–512. His students—now stewn across the world—remember him best for his virtuoso courses in topographic map interpretation, and in field geomorphology. Mackin simply demanded that his students use their eyes and attach them to their brains, and had no patience for those who refused to learn.

Pierre Dansereau, distinguished retired dean of the College of Science at the University of Montréal, and sometime Assistant Curator of the New York Botanical Garden, was fortunately on the botany faculty at the University of Michigan when I was a puling infant of a graduate student in 1951. He is internationally known for his works in ecology, botany, and metropolitan planning, but I knew him first in his course, "Vegetations of the World," in which I enrolled almost accidentally—surely one of the happiest accidents of my academic life. His lecture, comparing the look of English, French, Brazilian, and American gardens, and how those domestic gardens reflect national culture, caused scales to fall from my eyes in great clattering heaps. For all his prolific writing, he has unaccountably never written the lecture down for publication. Meantime, I take solace in two of Dansereau's brief essays, which reveal his remarkable perspicacity: "The Barefoot Scientist," *Colorado Quarterly* 12 (1962):101–15; and "New Zealand Revisited," *Garden Journal* 12 (1962):1–6. The latter essay, a combination of botanical observation, personal reflection, and acute social observation, reminds me delightedly of Charles Lyell's accounts of his travels in the American South in 1841–42. Both represent a tradition of scientist-cum-traveler, à la Teilhard de Chardin, that requires a combination of gentility and intelligence that is all too rare in our contemporary scientific community.

5. Sauer's influence was so pervasive and eclectic that it would be wrong to tag him with the label "landscape-reader." In fact, Sauer's impact on landscape-readers was felt more through the work of his students than through his voluminous writings. For an appreciation of Sauer as teacher see Dan Stanislawski, "Carl Ortwin Sauer, 1889–1975," *Journal of Geography* 74 (1975):548–54.

6. Professor Kniffen has been writing and teaching indefatiguably for forty years. Perhaps his most influential professional writing is "Folk Housing: Key to Diffusion," *Annals*, Association of American Geographers 55 (1965):549–77.

7. Like most members of the Berkeley school, Zelinsky is a prolific and eclectic writer, who is not easy to categorize. Some of his admirers would doubtless turn to *The Cultural Geography of the United States* (Englewood Cliffs, N.J.: Prentice-Hall, 1973), for an example of Zelinsky's breadth of interest. For Zelinsky-as-landscape-reader, however, I would nominate a lovely little thing, "Where the South Begins: The Northern Limit of the Cis-Appalachian South in Terms of Settlement Landscape," *Social Forces* 30 (1959):172–78.

8. Lowenthal is as prolific as Zelinsky. See "The American Scene," *Geographical Review*, 58 (1968):61–88, a thoughtful essay about why America looks the way it does and what that look has to say about our national character.

9. Parsons, one of Sauer's successors as chairman of the Geography Department at Berkeley, would doubtless disclaim the proud title of "landscape-reader," but I know better. He is—his students tell me—at his best with a small class, driving at breakneck speed along Interstate Five in the San Joaquin Valley, discoursing on the changing agricultural landscape that most travelers between Sacramento and Los Angeles find incomprehensible and (therefore) dull. For an eloquent statement of Parsons's views on "the significance and rewards of field observation," see "Geography as Exploration and Discovery," his presidential address to the Association of American Geographers, *Annals*, Association of American Geographers 67 (1977):1–16.

10. Hart's real passion is rural America, and his most comprehensive work on the subject is *The Look of the Land*, (Englewood Cliffs, N.J., Prentice-Hall, 1974).

11. Most "geographic field guides" are guaranteed to provoke instant ennui. For a splendid exception see Swain and Mather's *St. Croix Border Country* (Prescott, Wisc.: Pierce Country Geographical Society, 1968), an amusing and perceptive guide to the borderland of a growing metropolis.

12. Jackson founded the journal *Landscape* in 1951, and it was Jackson's personal testament until he retired in 1968. Those who love the American landscape and love trying to understand it owe him an incalculable debt, for *Landscape* helped teach a generation of neophyte geographers that there was nothing disreputable about going outdoors and asking naive questions about what one saw. More than a dozen of Jackson's wonderful essays are collected in *Landscapes: Selected Writings of J. B. Jackson*, ed., Ervin H. Zube (Amherst: University of Massachusetts Press, 1970). Some of Jackson's greatest coups, however, appear as unsigned notes and comments, published intermittently in *Landscape* during the halcyon days of his editorship in the 1950s and 1960s; see Meinig's review in the concluding essay of this book.

13. Glassie has the rare ability to combine painstaking research with lively writing. He has produced altogether some of the most interesting material in American folklore today. His most accessible work, and most sweeping in scope, is *Pattern in the Material Folk Culture of the Eastern United States* Philadelphia: University of Pennsylvania Press. Get the 1971 paperback edition; the 1968 hardback has no index and needs one badly.

14. George Stewart, *U.S. 40: Cross Section of the United States of America* (Boston: Houghton Mifflin, 1953), is one of the single best books about the United States. It is a combination of plain English and plain photography—breathtaking in clarity and content.

15. Wolfe made this remark in a public lecture at the Pennsylvania State University, University Park, in January 1976.

16. Alan Gowans, *Images of American Living: Four Centuries of Architecture and Furniture as Cultural Expression* (Philadelphia: Lippincott, 1964). Gowans is an architectural historian, and his book is a tour de force. Gowans was born a Canadian and, like many who are not American-born, sees the United States a good deal more clearly than most Americans do. Gowans has a similar book on Canada: *Building Canada: An Architectural History of Canadian Life* (Toronto: Oxford University Press, 1964). Reyner Banham, *Los Angeles: The Architecture of the Four Ecologies* (New York: Harper & Row, 1971), is the first and only book I have read that makes genuine sense of Los Angeles and why it looks the way it does. (For what it is worth, Banham's book made me shed all my academic-chic prejudices against Los Angeles, and look at that superb city with new eyes.) It is the best single book on an American city I have read. Banham, another architectural historian, is British, and his nationality forces one, I fear, to draw the same conclusion as one draws from the works and origin of Alan Gowans. Grady Clay (see Note 3) is *mirabile dictu*, American-born. Editor of *Landscape Architecture Quarterly*, Clay is both landscape architect and journalist, a happy combination.

17. Peirce Lewis, "Common Houses, Cultural Spoor," *Landscape* 19 (January 1975): 1–2.

18. Russell Lynes elaborates this idea with wit and intelligence in *The Tastemakers* (New York: Grosset and Dunlap, 1949).

19. Caution! If a scholar starts studying elephant-shaped hotels, he is likely to be denounced, or ridiculed, or pointedly ignored by self-styled "serious scholars," who will dismiss him as doing work that is "trivial" or "irrelevant." Students of landscape must learn to ignore such folk, or they will end up paranoid. The critics are more to be pitied than censured, since they see that other people are having fun, and they probably aren't. It's a sad fact—but nonetheless a fact—that many academics don't consider it respectable to enjoy their scholarship. But life is too short to worry about people like that, so keep your eyes open, and remember that you're trying to understand nothing less serious (or less funny) than American culture.

20. The volume of printed material about the Brooklyn Bridge must equal that of the bridge itself. The beginner might try David McCullough, *The Great Bridge: The Epic Story of the Building of the Brooklyn Bridge* (New York: Simon & Schuster,

1972). McCullough's book is a panoramic portrait of American life during the two decades the bridge was abuilding: economic, technology, politics, social mores, urban history—almost nothing is not touched on. It is, in fact, a kind of cultural ecology of the bridge.

Bridges, of course, are mighty symbols in this mobile society. The best general work is David Plowden, *Bridges: The Spans of North America* (New York: Viking Press, 1974), technically irreproachable, and undergirded with the kinds of photographs that have made Plowden justly famous as one of the most sensitive contemporary photographers in America. If we had an encyclopaedia of volumes to match Plowden's on other topics in the vernacular landscape there would be no need for me or anyone else to complain about inattention to ordinary elements of ordinary landscapes.

21. Scholars have been complaining about such neglect for a long time. See, e.g., Reyner Banham "The Missing Motel," *Landscape* 15 (1965):4–6. Nevertheless, scholars continue to chip away at the monolithic indifference. The Winter 1976 issue of *Landscape* contains a fine succinct essay on the origin and evolution of garages and carports in the U. S. by J. B. Jackson.

22. Afficionados will differ on what constitutes real vintage Wolfe. For Wolfe-as-landscape-analyst, my preference is the essay on Las Vegas signs in *The Kandy-Kolored Tangerine-Flake Streamline Baby* (New York: Farrar, Straus, & Giroux, 1965). For Wolfe-as-social-observer, he reaches Mark Twainian levels in his *Radical Chic and Mau-Mauing the Flak-Catchers* (New York: Farrar, Straus, & Giroux, 1970).

23. *Ulrich's International Periodicals Dictionary* lists a bewildering variety of journals—trade and otherwise. In some instances, however, a bit of legwork will yield better results. If you want *the* authoritative trade journal on, say, the motel industry, just drop by the nearest prosperous-looking motel and ask the manager if you can see his copy. You can be sure he subscribes, and equally sure that he will talk your ear off about motel management, if he has the time.

24. Melvin E. Hecht, 'The Decline of the Grass Lawn Tradition in Tucson," *Landscape* 19 (1975):3–10.

25. See Note 14.

26. See Note 3.

27. The essay is reprinted in Zube, *Landscapes*.

28. For an astonishing variety of information about American technology and its social context, see J.C. Furnas, *The Americans: A Social History of the U.S., 1589–1914* (New York: Putnam's 1969). Furnas covers a wonderful range of topics, cites a formidable bibliography and, unlike the standard history of technology with its concern with inventors, pays special heed to the effects of technological innovation, and the social effects of successful invention.

29. David Lowenthal, "The American Way of History," *Columbia Univ. Forum* 9 (1966):27–32; "Past Time, Present Place: Landscape and Memory," *Geographical Review* 65 (1975):1–36; Peirce Lewis "The Future of the Past: Our Clouded View of Historic Preservation," *Pioneer America* 7 (1975):1–20.

30. But see, again, Furnas *The Americans* and Zube *Landscapes*.

31. The rewards can be greatly multiplied if one draws pictures of what one sees. I do not mean arty impressionistic sketches; I mean literal, primitive drawings, the virtue of which is to *force* one to notice details that might otherwise go unseen. A similar device (which will to cause artists to recoil in disgust) is to project a slide onto a piece of paper and draw the image, omitting as little as one can. Mere tracery? Certainly. Cheating? Certainly not. One is learning to look and see details, not to render masterpieces.

32. D. W. Meinig. "Environmental Appreciation: Localities as a Humane Art," *Western Humanities Review*, 25 (1971):1–11.

The Beholding Eye

Ten Versions of the Same Scene

D.W. Meinig

"Landscape" is at once an old and pleasant word in common speech and a technical term in special professions.[1] As Americans become more conscious of and concerned about their visible surroundings—their environment—it is going to crop up more frequently in both realms of conversation and it may be useful occasionally to consider a difficulty that almost inevitably arises as soon as we attempt to communicate beyond very narrow professional circles.

A simple exercise will quickly reveal the problem. Take a small but varied company to any convenient viewing place overlooking some portion of city and countryside and have each, in turn, describe the "landscape" (that "stretch of country as seen from a single point," as the dictionary defines it), to detail what it is composed of and say something about the "meaning" of what can be seen. It will soon be apparent that even though we gather together and look in the same direction at the same instant, we will not—we cannot—see the same landscape. We may certainly agree that we will see many of the same elements—houses, roads, trees, hills—in terms of such denotations as number, form, dimension, and color, but such facts take on meaning only through association; they must be fitted together according to some coherent body of ideas. Thus we confront the central problem: any landscape is composed not only of what lies before our eyes but what lies within our heads.

Recognition of that fact brings us to the brink of some formidably complex matters. But it is not necessary to plunge into the technical thickets of optics, psychology, epistemology, or culture to converse intelligently about the topic. It is far too fascinating and important to be left fragmented and obscured in the jargon of such specialists. It deserves the broad attention that only ordinary language allows. And so let us review some of the different ways our varied group might describe a common scene. We are concerned not with the elements but with the essence, with the organizing ideas we use to make sense out of what we see.

There are those who look out upon that variegated scene and see, first and last,

landscape as Nature.

For them all the works of man are paltry compared with nature, which is primary, fundamental, dominant, enduring. The "vault of heaven," the "rock of ages," the "everlasting hills," are old metaphors which tell us that if we really ponder the landscape, it is nature that controls. The sky above, the ground beneath, and the horizon binding the two provide the basic frame, holding within the lay of the land, its contours and textures; the weather and the light, ever-changing with the hours and seasons, affecting all our perceptions; and at all times some display of the power of nature, its quiet inexorable rhythms, the power of growth, of moving water, the immense power of storms. Amidst all this man is miniscule, surficial, ephemeral, subordinate. Whatever he does upon the surface of the earth, even his greatest skyscrapers, dams, and bridges, are, by comparison, minute, feeble, and transitory; mere scratchings on the skin of Mother Earth.

Such a viewer is ever tempted in his mind's eye to remove man from the scene, to restore nature to her pristine condition, to reclothe the hills with the primeval forest, clear off the settlements, heal the wounds and mend the natural fabric—to imagine what the area is *really* like. It is an old and deeply rooted view which separates man and nature. Ideologically it had its greatest vogue in eighteenth century Romanticism, in that longing for wilderness, in the view of nature as pure, fine, good, truly beautiful. It had a major impact upon nineteenth century science, as the very term "natural sciences" attests.

It can be a seductive view. It is not hard to see beauty and power in nature. One can feel an awe and majesty even in mere depictions of nature, as in the photgraphs of Ansel Adams and those beautiful books of the Sierra Club. And it is a view which may again become more common, for the more people begin to see man's works as despoliation, the more they will see pristine nature as perfection, as a baseline from which to measure corruption.

The romantic view is in fact very much alive, usually, perhaps necessarily, expressed as a kind of nostalgia:

> There was a time, in the sweet childhood of the human race, when man lived close to nature ... the world of nature and the world of man were synonymous. ... [2]

But that describes a unity rather than a separation, and it is quite possible even today to regard

landscape as Habitat.

In such a view, every landscape is a piece of the Earth as the Home of Man. What we see before us is man continuously working at a viable relationship with nature, adapting to major features, altering in productive ways, creating resources out of nature's materials; in short, man domesticating the earth.

The basic patterns in the landscape, the patchwork of fields, pasture and woods, of homesteads and villages, the plan of cities and suburbs, all reveal man's conscious selection of soils and slopes, elevations and exposures, sites and routes provided in the beginning by nature. So too the very shapes, colors, textures, and other qualities of things, of fences and buildings, of trees and flowers, animals and birds, reflect man's selection from earth's great bounty and his reworking, retraining, rearranging into desirable forms. And man himself in so many ways, in diet and dress, emblems and rituals, in his everyday work and play, reveals his adaptations, often subtly and unconsciously, to nature.

Every landscape is therefore basically a blend of man and nature. Man may make mistakes, damage nature and thereby himself, but in the long run man learns and nature heals. Thus even when landscape seems to display some maladjustment, it is only a phase in man the domesticate working toward symbiosis, a process he has been engaged in for a million years.

This, too, is an old and attractive view. It is the ideology of the harmony of man and nature, of the earth as the garden of mankind, of man as the steward, the caretaker, the cultivator. Man must adjust to nature, but nature is basically benign and good and when properly understood will provide a comfortable and enduring home. It is a view never better expressed than in Ellen Churchill Semple's opening lines sixty-five years ago in her monumental *Influences of Geographic Environment:*

> man is a child of the earth, dust of her dust; the earth has mothered him, fed him, set him tasks, directed his thoughts, confronted him with difficulties ... given him problems ... and at the same time whispered hints of their solution. ... [3]

It is an ideology which had a major impact upon a number of fields, especially upon the early stages of human ecology and anthropogeography. The central working concept was "environmentalism" in one form or another. It strongly shaped those classic regional monographs in France, a rich body of rural studies in Europe, and underlies admiration for the richly humanized landscapes of the peasant world. Until recently China's "farmers of forty centuries" were often cited as a model of harmonious adaptation and the Jeffersonian yeoman farmer is one among many related idealizations in Western thought.

The general concept is not only still alive, it is rapidly gathering strength in somewhat more sophisticated form. It lurks in various guises within much of the recent literature on ecology and environment. But as man's power to affect the earth has increased, his reworking of nature may appear to be less an adjustment and more so fundamental an alteration that one may see the

landscape as Artifact.

Such a person sees first of all and everywhere the mark of man in everything. Nature is fundamental only in a simple literal sense: nature provides a stage. The earth is a platform, but all thereon is furnished with man's effects so extensively that you cannot find a scrap of pristine nature. The soils, trees, and streams are not "nature" as distinct from man, they are profoundly human creations: soils altered by plowing, cropping, burning, mulching, fertilizing, draining; forests cut and burned and the whole complex changed by new associations of species; streams silted, channeled, their regime affected by myriad changes in their watersheds. The very shape of the land surface has been modified in a thousand ways, by cuts and quarries, excavations and embankments, fills, dams, culverts, terraces, revetments. Even the weather, and especially that most directly affecting man, near the ground, has been altered by changes in surfaces and in the heat, dust, and chemicals discharged into the air. But also the weather is no longer very important, for man lives increasingly indoors in carefully controlled atmospheres.

In this view it is thus idle sentiment to talk of man adapting to nature in modern America. Indeed his buildings and streets and highways appear more often to be sited in utter disregard for the contours of nature. A rigid linear geometry has been set discordantly but relentlessly upon the varied curves of nature. So comprehensive and powerful has been man's role in changing the face of the earth that the whole landscape has become an artifact.

Ideologically this is a view of man as creator, not only emancipated from, but the conqueror of, nature. Although the concept may have roots deep in history, its full flowering is recent. In science it is marked by recognition of man as ecologically dominant. The work of George Perkins Marsh more than a century ago is an early landmark in calling attention to man's impact,[4] but the twentieth century concept of man as technocrat in charge of remolding the earth to suit his desires marks the more radical shift. It is concomitant with the growth in the pervasive power of the engineer to alter the physical earth and of the biologist to alter organic life.

But the motivation of science is deeper than this utilitarian, manipulative expression. For the scientist, driven by a desire for understanding for its own sake, engaged in the endless exploration of the world we live in, may look out upon our scene and see

landscape as System.

He may see all that lies before his eyes as an immense and intricate system of systems. The land, the trees, roads, buildings, and man are regarded not as individual objects, ensembles of varied elements, or classes of phenomena, but as surficial clues of underlying processes. Such a mind sees a river not as a river, but as a link in the hydrologic circuit, a medium of transport carrying certain volumes of material at a certain rate within a segment of a cycle, a force altering the shape of land in a consistent calculable way. Such a mind sees trees not in terms of species, dimension, color, nor even as major organic features, but as chemical factories powered by sunlight, lifting stations in the hydrologic cycle, biological transformers in the energy exchange between lithosphere and atmosphere. In such a view landscape is a dynamic equilibrium of interacting processes.

Man is of course an inexorable part of these systems in one way or another. His more obvious structures and movements in the landscape are most likely to be seen as "functions," that is, as processes undertaken for rational purposes. Houses, garages, barns, offices, stores, factories are all "service stations" and "transformers," and may be regarded as crude, imperfect, outward expressions of abstract social and economic systems.

Such a view is wholly the product of science, a means of looking inside matter to understand things not apparent to the naked untrained eye. It is a view still in vigorous development, beginning with analysis, disintegrating things into their parts, and turning increasingly to synthesis, putting things together in such a way as to give us a new level of understanding interrelationships. It is likewise the view of social science, which seeks to emulate physical science, and finds its reality not in persons or idiosyncratic arts, but in aggregates, in group behavior.

For such persons the landscape that others may see is only a facade which their vision penetrates to reveal a transect of intricate pulsating networks, flows, interactions, an immense input-output matrix. To the extent that it can be understood, it takes on "reality" for them in diagrams, schemata, formulae. It is an ideology that implies a faith in man as essentially omniscient; that man through the rigorous disciplined power of his mind will eventually understand all that lies before him in the landscape; that ultimately through science we shall know the truth.

Of course we are far from knowing enough as yet and thus the landscape can be regarded as a laboratory, an experiment station. Actually, because science, by its nature, demands intense specialization from most of its practitioners, no one viewer can envision anything like the full range of questions to be asked, and no landscape will serve as an equally suitable laboratory for the full range of specialists. But the eyes of the fluvial geomorphologist and of the social psychologist have a similar kind of selectiv-

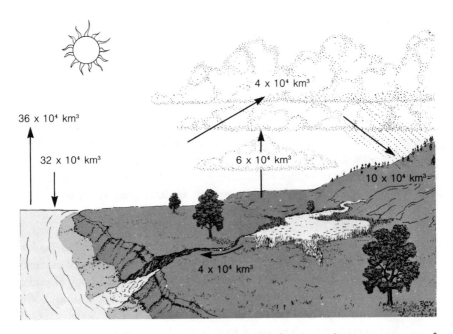

To see landscape as a system is to penetrate the facade to discern a transect of pulsating flows, an immense energy exchange: The hydrologic cycle. (Leo Laporte, *Encounter With The Earth* [San Francisco: Canfield Press, 1975])

ity, seeking the general amidst the particular, to build similar kinds of abstractions for similar purposes. Because any such findings need repeated testing, any one landscape can be no more than a sample area.

Such may be the way in which the basic research scientist regards our scene, but there are others who may be armed with similar tools but see it very differently, for they see every

landscape as Problem.

That is, see it not as a problem in the scientific sense of a need to know more in order to understand better, but as a condition needing correction.

To such a person the evidence looms in most any view: eroded hills, flooding rivers, shattered woodlands, dying trees, dilapidated farms, industrial pollution, urban sprawl, neon strips; garbage and grit, smog and sewage, congestion and clutter, and amidst it all, people impoverished in body or spirit. For such a person, other views of landscape are utterly inadequate. To regard the scene before us as no more than a laboratory for so-called objective research is to be indifferent to human needs; every landscape evokes wrath and alarm, it is a mirror of the ills of our society and cries out for drastic change.

Nevertheless, this view of landscape through the eyes of the social actionist may incorporate something from all these other views: it evokes a reverence for nature, a deeply felt concern for the earth as habitat, and a conviction that we have the scientific ability to right these wrongs. What is needed is a far greater awareness of what is happening and why. It is thus a view which tends toward a humanism harnessed to politics in the hope of generating a genuine populist movement against what is regarded as a callous, selfish, or simply inert establishment.

Perhaps the basic scripture of this movement is that masterpiece of quiet horror, Rachael Carson's *Silent Spring*. It is an apocalypse, a Book of Revelation of the last days of life on earth. But the most powerful evidence is the landscape itself and thus the most effective tract is a book of photographs showing what desecration we have wrought, as, for example, William Bronson's display of California in *How to Kill a Golden State*.

But those who exhort us to look with alarm and act with whatever political clout we can muster represent only the more extremist wing of the viewers of landscape as problem. There is another set (in fact they overlap a good deal) which is not so much a shrill citizenry as an interrelated group of professions for whom every landscape is a *design problem*. The problems they see may be functional (congestion, danger, incompatible uses), aesthetic (clutter, lack of proportion), or something of both; their common perspective is to look at the landscape and imagine a different one: one they have redesigned. It is not that every landscape is in crisis, but that every one is a challenge, every landscape induces a strong itch to alter it in some way so as to bring about a more pleasing harmony and efficiency.

Ideologically, such persons are expressing a strong humanism grounded in science and linked to aesthetics which seeks to apply professional skills to making over the earth. It is obviously closely related to the view of landscape as artifact: the critical difference lies in the realm of control and comprehensive planning. The title of a well-known book expresses it succinctly: *Man-made America: Chaos or Control.*[5] And therefrom arises a whole set of grave problems for any democratic society: who is to control? by what means? to what extent? for what purpose? (And therein lies ample justification for ever-wider discussions of "landscape" at every level.) Whereas the tool of the social actionist is the propaganda tract featuring the worst to be seen in the real landscape, that of the designer is the plot plan, the sketch, the perspective of the imagined landscape when improved by his application of art and technology.

Such design specialists are not alone in imagining "improved landscapes." They are in fact far outnumbered by those who see

landscape as Wealth.

Such persons are wont to look upon every scene with the eyes of an appraiser, assigning a monetary value to everything in view.

It is a comprehensive view, for everything has or affects value within a market economy. And it is a logical and systematic view which is continually adjusted to keep it concordant with ever-changing reality, for appraisals of property are recurrently tested by actual transactions which affect not only that sold but others adjacent or of similar kind. Like that of science, it is a penetrating view which looks beyond the facade to peer within and to organize what it finds in abstractions. It looks at a house and sees square-footage and the number of bedrooms and bathrooms; it looks at a business building and sees length of frontage, capacity, storage space, delivery access. It is a keen geographical view which reflects a quick sense of how things are actually arranged in a landscape, for relative location, quality of neighborhood, and accessibility are fundamental determinants of value. It takes note of age, but with a concern for depreciation, obsolescence, fashion, prestige, rather than an interest in history as such.

Public properties—schools, libraries, streets, parks, reservoirs, garbage dumps—are fitted into the system, for each affects the value of its surroundings, as do other site qualities—trees, hills, valleys, and especially "views" from residential property. Furthermore, people have a place in such an appraisal, for where the rich and where the poor live, work, shop, play, and go to school greatly affect property values.

Such a view of landscape is future-oriented, for market values are always undergoing change and one must assess their trends. Such, obviously, is the view of the speculator, but it is also the view of the developer and is thus akin to that of the landscape designer, for "development" is usually thought of as "improvement" and may involve strong feelings of creativity and of contributing to the benefit of society. The fact that it also enhances the developer's personal wealth taints it with selfishness, but vanity may also have a shaping influence upon the designs of the planner and landscape architect, and we should be wary of making invidious distinctions.

This view of landscape as wealth is of course strongly-rooted in American ideology and reflective of our cultural values. It represents our general acceptance of the idea that land is primarily a form of capital and only secondarily home or familial inheritance; that all land, all resources, are for sale at any time if the price is right; that speculation in land is a time-honored way to wealth.

Such a view is clearly the mark of a society which is strongly commercial, dynamic, pragmatic, quantitative in its thinking and the very landscape itself must reflect such characteristics. So much so, that one can sit upon that hilltop, look out over our scene, and see

landscape as Ideology.

Just as the scientist looks through the facade of obvious elements and sees processes in operation, so others may see those same elements as clues and the whole scene as a symbol of the values, the governing ideas, the underlying philosophies of a culture. Where those who see landscape as problem see disorder, clutter, incongruity, congestion, pollution, sprawl, and dereliction amid the glitter, those who see it as ideology may see distinct manifestations of American interpretations of freedom, individualism, competition, utility, power, modernity, expansion, progress. That does not mean they cannot see the problems, but that they are more concerned to look more deeply to see how the landscape represents a translation of philosophy into tangible features.

For such persons a quiet, reflective study of an American landscape may evoke not only those ideas but the men associated with them, so that, hovering like ghosts over the distant view, are its real creators. Who are they? No two viewers are likely to visualize the same pantheon, but John Locke, Adam Smith, Charles Darwin, Thomas Jefferson, Frederick Jackson Turner, John Dewey might stand out fairly clearly.

To see landscapes in such terms is to see as a social philosopher and to express a firm belief that broad philosophical ideas matter in very specific ways. It is a view which clearly insists that if we want to change the landscape in important ways we shall have to change the ideas that have created and sustained what we see. And the landscape so vividly reflects really fundamental ideas that such change requires far-reaching alterations in the social system. Hence, for example, the scorn for "beautification"—the planting of flowers by the roadside—as a mere cosmetic which masks the need for painful change.

To see landscape as ideology is to think about how it was created, but there is another way of doing that which, while at its best is reflective and philosophic, is also much more detailed and concrete: to see

landscape as History.

To such a viewer all that lies before his eyes is a complex cumulative record of the work of nature and man in this particular place. In its most inclusive form it sends the mind back through the written record and deep into natural history and geology. More commonly it reaches only back to early man, and usually in America to the first European settlers.

The principal organizing system is chronology, which is not in itself history but the scaffold upon which one constructs history. Thus every object must be dated as to origin and to significant subsequent change. Exact dating may require tedious research, but the skilled landscape historian working in a generally familiar culture area can assign approximate dates to most items based upon materials, design, ornamentation, purpose, position. By classifying features according to age the landscape can be visualized in terms of layers of history, which are sometimes rather distinctly separated in area, as with a new housing tract, but more often complexly interwoven.

The visible landscape is not a full record of history, but it will yield to diligence and inference a great deal more than meets the casual eye. The historian becomes a skilled detective reconstructing from all sorts of bits and pieces the patterns of the past. He learns how indelible certain features tend to be, such as the basic geometry of routes and lots, and how changeable and deceptive others are, such as facades and functions. And there is much more to be learned than chronological changes. The physiognomy of a house, its size, shape, material, decoration, yard, outbuildings, and position, tells us something about the way people lived. Furthermore every house had its particular builder and each has been lived in by particular individuals and families and something of that, too, may perhaps be read in the landscape.

This can be a view of landscape as process, but with a different emphasis from that of the scientist. Where the latter sees an association of classes of things being affected by generalized processes to form a general

To see landscape as ideology may lead to attempts to envision the persons who have shaped the underlying philosophies of a culture. The citizens of Bukhara have been provided with a rather explicit visual aid. (R.G. Jensen)

pattern of predictable events, the historian sees the particular cumulative effects of processes working upon the particular elements of this locality. The degree to which the historian relates the particular to the general depends upon his purpose, but any historical view clearly implies a belief that the past has fundamental significance, one aspect of which is so pervasive as to be easily overlooked: the powerful fact that life must be lived amidst that which was made before. Every landscape is an accumulation. The past endures; the imprint of distant forebears in survey lines, land parcels, political jurisdictions, and routeways may form a relatively rigid matrix even in areas of rapid change. The landscape is an enormously rich store of data about the peoples and societies which have created it, but such data must be placed in its appropriate historic context if it is to be interpreted correctly. So, too, the landscape is a great exhibit of consequences, although the links between specific attitudes, decisions, actions, and specific results may be difficult to trace with assurance. In any case, whether the historical view is meant to serve curiosity, reflection, or instruction, the landscape provides infinite possibilities.

There is a logical complement to this view of landscape as history, one which overlaps and is yet distinct in perspective and purpose: a view of

landscape as Place.

In this view every landscape is a locality, an individual piece in the infinitely varied mosaic of the earth. Such a viewer begins by being at once comprehensive and naive: by encompassing all and accepting everything he sees as being of some interest. It is landscape as environment, embracing all that we live amidst, and thus it cultivates a sensitivity to detail, to texture, color, all the nuances of visual relationships, and more, for environment engages all of our senses, the sounds and smells and ineffable feel of a place as well. Such a viewer attempts to penetrate common generalizations to appreciate the unique flavor of whatever he encounters.

It is the view cultivated by serious travel writers with the effective assistance of the photograph and the sketch to display both physiognomy and impressions of a place. Closely akin, with a greater emphasis upon individual persons in their environments, is the work of the "local" or "regional" novelists, the best of whom can evoke a keen sense of the individuality of places.

Such a view is also old and central ground to the geographer, whose field has at times been defined as a study of the characteristics of places. The chief badge of the geographer is the map. To him a place is at once a location, an environment, and an areal composition, and the last is best expressed on a map, a symbolization of the spatial arrangement of the elements of the locality. Compositions have form, and the geographer will see in the landscape a variety of areal patterns and relationships: clusters, nodes, scatterings, gradations, mixtures. These of course take on meaning only when interpreted with some understanding of history and ideology, of processes, functions, and behavior, and of larger geographic contexts. And the geographer, like the historian, can pursue his interests in either direction: toward generalization or particularity.

Those interested in particular localities share a belief that one of the greatest riches of the earth is its immense variety of places. It is a view which far transcends the banal tourist search for the exotic; true believers are comprehensive: literally every place is of some interest. Indeed, it is a view which suggests that a well-cultivated sense of place is an important dimension of human well-being. Carried further, one may discover an implicit ideology that the individuality of places is a fundamental characteristic of subtle and immense importance to life on earth, that all human events *take place*, all problems are anchored in place, and ultimately can only be understood in such terms. Such a view insists that our individual lives are necessarily affected in myriad ways by the particular localities in which we live, that it is simply inconceivable that anyone could be the same person in a different place.

This opening of one's senses to "get the feel" of a place is close to still another view

landscape as Aesthetic.

There are many levels and varieties to this view, but all have in common a subordination of any interest in the identity and function of specific features to a preoccupation with their artistic qualities.

"Artistic quality" is of course a matter of endless controversy. It is well known that landscape painting or drawing as a special genre is a feature peculiar to certain eras of certain cultures. The very idea of landscape as scenery is a surprisingly late development in Western culture, requiring as it does a special conscious detachment by the observer. Within the realm of landscape painting we will find examples which express many of the views of landscape discussed: the power and majesty of nature, the harmony of man and nature, the mark of history upon the land, the detailed character of places. Each of these represents a careful selection by the artist. But the "purest" form of landscape as aesthetic is a more comprehensive abstraction in which all specific forms are dissolved into the basic language of art: into color, texture, mass, line, position, symmetry, balance, tension. The versions and variations are infinite in this most individualistic view of landscape.

This, too, is a penetrating view. It seeks a meaning which is not explicit in the ordinary forms. It rests upon the belief that there is something close to the essence, to beauty and truth, in the landscape. Landscape becomes a mystery holding meanings we strive to grasp but cannot reach, and the artist is a kind of gnostic delving into these mysteries in his own private ways but trying to take us with him and to show what he has

Landscape as aesthetic in which the specific forms begin to dissolve into the basic language of art: Charles Sheeler, *Midwest*, 1954. (Walker Art Center, Minneapolis)

found. In this view landscape lies utterly beyond science, holding meanings which link us as individual souls and psyches to an ineffable and infinite world.

Ten landscapes do not exhaust the possibilities of such a scene, but they do suggest something of the complexities of the topic. Identification of these different bases for the variations in interpretations of what we see is a step toward more effective communication. For those of us who are convinced that landscapes mirror and landscapes matter, that they tell us much about the values we hold and at the same time affect the quality of the lives we lead, there is ever the need for wider conversations about ideas and impressions and concerns relating to the landscapes we share.[6]

Notes

1. Marvin Mikesell, "Landscape," *International Encyclopedia of the Social Sciences*, vol. 8, (New York: Crowell-Collier and Macmillan, 1968), pp. 575–80.
2. Garrett Eckbo, *The Landscape We See* (New York: McGraw-Hill, 1969), p. 42.
3. Ellen Churchill Semple, *Influences of Geographic Environment* (New York: Henry Holt, 1911), p. 1.

4. George Perkins Marsh, *Man and Nature; or, Physical Geography as Modified by Human Action* (New York: Scribner, 1864; reprinted in the John Harvard Library, Cambridge: Harvard University Press, 1965).

5. Christopher Tunnard and Boris Pushkarev, *Man-made America: Chaos or Control?* (New Haven and London: Yale University Press, 1963).

6. For more explicit suggestions see D. W. Meinig, "Environmental Appreciation: Localities as a Humane Art," *The Western Humanities Review* 25 (1971):1–11.

II

Explorations

The Biography of Landscape

Cause and Culpability

Marwyn S. Samuels

> In every action what is primarily intended by the
> doer, whether he acts from natural necessity or
> free will, is the disclosure of his own image . . .
> Nothing acts unless it makes patent its latent self.
>
> *Dante*

In *The Making of the English Landscape* Hoskins reports that on January 19, 1439 a certain Thomas Flower failed to repair a section of the Wisbeach fendike.[1] The result was that flood waters pouring down from the Midlands broke through the dikes and inundated some thirteen thousand acres of land. By itself, the event seems hardly worth reporting. Farmers the world over have frequently neglected their fields and dikes, and often with spectacular consequence. The only thing remarkable about the case of Mr. Flower is that we somehow know his name and the date of his folly. What makes that remarkable is that Mr. Flower, a yeoman farmer like countless others, represents what we and most historians would consider a "nobody in particular." As Hoskins would have put it, he was one of those farmers "whose name may be made known to us if we search diligently enough."[2]

And yet, so what? Assuming that with the greatest diligence we are able to uncover the names of any number of otherwise obscure farmers and peasants the world over, what possible significance could that have? Even if the event had left an indelible imprint on the landscape, what good would it do us now to know who to blame? In searching out the "cause" for Mr. Flower's recklessness, chances are that all we would find would be some prosaic fact regarding his lack of character and responsibility.

Why then should Hoskins have bothered to tell us about his failure? Equally, as he does elsewhere, why should he have bothered to inform us about Mr. Lancelot (alias Capability) Brown who in the 18th century designed Kew Gardens and Blenheim Palace?[3] If our interest is in the landscape, what difference does it make who occupied, designed, or shaped the landscape? The thing is, it represents a certain pattern, style or motif that emerged in the wake of other patterns, styles and motifs. We can trace its aesthetic and institutional origins and be satisfied that it "derived" under the influence of Chinoisere and Physiocratic idealism. Or, we can assign the landscape to various economic, social, political and broadly cultural "forces." We can assign its "underlying impetus" to such "processes" as the industrial revolution, the spirit of capitalism, the doctrine of progress, or to the "nature" of *homo economicus, homo politicus, homo laborens, homo ludens,* and of course, *homo sapien.* We can, as it were, *explain* the landscape without so much as a passing reference to anyone in particular who happened to live in, pass through, influence, or even make the landscape. All such individuals become "meaningless" as they are explained away in the wake of one or another all-encompassing "process" that "alone makes meaningful whatever it happens to carry along."[4]

Yet there is something vaguely amiss here. However rational, there is something unreasonable about a human landscape lacking in inhabitants; something strangely absurd about a geography of man devoid of men. The fact that we need not bother identify anyone in particular, that culpability and responsibility in and for the landscape have become irrelevant to the quest for landscape meaning, moreover, reveals something terribly wrong about the way in which we look at the event and assess the meaning of landscape. It betrays an intellectual and perhaps a cultural milieu in which the most ordinary of questions—who did it—is curiously irrelevant to the meanings we give our landscapes. Simply phrased, it unveils a context in which the idiosyncratic, the particular, the individual himself and the self itself have lost much of their own meaning.

How is it that the *who* behind the image and facts of landscape, or the "biography of landscape," no longer grips our attentions? At the same time, why is it that even as we explain away the landscape, some vestige of lost self remains both to nudge our conscience and demand a search for some Mr. Flower or Mr. Brown?

The Context of Latent Selfhood

The origins of lost individuation lie deeply imbedded in the intellectual past of the West. It is perhaps only a slight metaphorical exaggeration to suggest that those origins are traced by an intellectual map formed as a series of concentric zones pivoted on two cities: Athens and Jerusalem. Shaped on the one hand by the Parthenon of enlightened rationalism where "necessity knows no persuasion," and on the other by the rock of Abraham's trial where choice knows no refuge, the Western intellect has been torn between the determination of cause (explanation) and the assignment of responsibility (attribution). Where the one has left its legacy in the tranquil certainty of logical rectitude, the other has bequeathed the confusion and *angst* of uncertain responsibility. Where Athens delivered Aristotle, Descartes, Newton, Spinoza, Kant, Hegel, Marx, and more recently Carnap, Ayer, Wittgenstein and Russell, Jerusalem provided Job, the figure of Christ, Shakespeare, Cervantes, Goethe, Kierkegaard, Tolstoy, Dostoevsky, and more recently Shestov, Buber, Camus and Sartre. With the one we have been left with Spinoza's famous dictum: *non ridere, non lugere, neque destastari, sed intelligere* (not to laugh, not to lament, not to curse, but to understand). With the other we constantly engage lamentation, the passion of guilt, and the anguish of freedom. The result has been an intellectual milieu and a logical dilemma of two different, if sometimes overlapping, sensitivities.[5]

If this distinction bears the mark of some overstatement, it nonetheless has the merit of drawing our attention to what has long since become a critical problem for the Western intellect. The problem is simply this: Athens has overwhelmed Jerusalem, rendering responsibility a matter of explanation or reducing it to little more than a lingering doubt. That conquest goes by many names: "objectification," the ascent of the secular, and "alienation" among them; but it was perhaps best and most generically summed up by Lev Shestov as an unrelenting "struggle against the 'I,' against individual existence" which arises from the all too human endeavor to escape the particular and the consequences of selfhood. In this struggle against individual existence, *Unity* waged war and, as Shestov put it, has had "for its goal the destruction in this life of the 'hated I' which had escaped from the lap of the universal . . . (into) the realm of incomprehensible, painful, terrible, and consequently unacceptable freedom."[6] Confronted as much by the rationality of medieval Christian thought as by the pagan Reason of enlightened science, the enemy is the ego, the self, and even the soul, all of which find their integrity simply as dependent variables.

The War Against the Self: Medieval Contexts

The history of the war against the self is too rich to be given adequate treatment here, but it does bear importantly on the question of man's relationship to nature and thence to the meaning of landscape. It was, for example, in this context that St. Anselm in the twelfth century inveighed against the temptations of nature and especially against the charms of gardened landscapes. He did so because he, like virtually all theologians of the times, held that the senses—the datum of individuality, the self and the will—titillated by nature were positively sinful.[7] As St. Augustine had made clear, it was not nature that was evil or sinful, but man who, by virtue of the self, is "drunk with the invisible wine of (his) own perverted earth-bound will."[8] Indeed, one can trace this sentiment to the tale of man's fall from grace. Possessed of the fruit of selfhood, man is cast out of the gardened landscape for an act of will.[9]

This is not to suggest that the medieval Christian world was wholly antagonistic to the self and individuality. However "perverted," the self's "own will" was the essential premise of the Jewish and Christian concept of man, for the message of man's fall from grace was also that he alone of G-d's creations was possessed of will and conditioned by responsibility. Indeed, though fundamentally evil, the self remained the necessary prerequisite to the Christian doctrine of sin. What is more, the medieval world was not wholly lacking in a few lonely advocates for the self. In the fourteenth century, for example, there was Petrarch who, as Kenneth Clark noted, was "probably the first (Western) man to express an emotion on which the existence of landscape painting so largely depends."[10] Clark to the contrary notwithstanding, that emotion had little to do with either Petrarch's disdain for urban life or his adoration of nature. The medieval

Christian world abounds with examples of those, like St. Bernard and St. Francis, who adored nature as the work of G-d.[11] The monastic movement, furthermore, strongly evinced a disdain for urban life. The innovation that underscored Petrarch's deviation from the medieval norm was instead a forceful introspection gained through the act of observing nature. As he put the matter himself, having obtained the vista from Mount Ventoux, it then occurred to him that he might "long ago have learned . . . that nothing is wonderful but the soul, which, when great itself, finds nothing great outside itself." Satisfied that he had seen enough of the mountain view, he could then turn his "inward eye" upon himself and hike homeward in the quiet silence of introspection.[12]

In this act of introspection, and the assertation of a soul magnified beyond anything itself, Petrarch was perhaps, as Clark suggests, "the first modern man" who reminds us of ourselves "in his curiosity, his scepticism, his restlessness, his ambition, and his self-consciousness."[13] Yet he was not alone in this regard. The theme of introspection amidst the solitude of nature lies partly at the root of the institutions of hermitage and the cloister. Though mitigated by monastic orderliness and the inclination to manicure nature through monastic agriculture, the hermitage and cloisture also evidenced a proclivity for solitary introspection away from the social order of the city. The desire for a retreat from the city in order to better facilitate the confrontation with self and the relationship with G-d is deeply rooted in both East and West. Its Christian precedent is, of course, Jesus' own retreat into the Judean desert where, alone and away from the civilized temptations of the city, he was better able to face both G-d and himself, which is to say his responsibility as the "Christ." For this reason St. Jerome would later proclaim that "the city is a prison, the desert *loneliness* a paradise." For this reason too the history of asceticism (in the West and the East) invariably entails a geography of solitude, the landscape of which is devoid of humanity-in-general, but amidst nature, focused on the individual alone.

Despite such early precedents, however, Petrarch's revolution was largely stillborn, just as the loneliness of the cloister was overcome by the social order of the monastary.[14] The medieval world was almost everywhere conditioned by the desire for Order in nature and society unblemished by the ignominy of "perverted" self-centeredness and the will. Not even the capricious godhead was exempted from Unity, for as St. Thomas Aquinas put it, "only that is excluded from the divine omnipotence which contradicts the reason or essence of being. . . ."[15] The epitome and the symbolic end of the medieval world view are both probably nowhere better illustrated than in its case against Galileo. In its defense of the Ptolemaic system and its attack on the Galilean demonstration of Copernican theory, the medieval world betrayed an essential distrust of the senses and their "perverted" will. What Galileo did was to demonstrate the Archimedean point "with the certainty of sense perception" and it was that *empirical demonstration* rather than the Copernican *theory* that the Church found utterly threatening.[16]

Science and The Self: The esprit de systeme

Born of Galileo's empiricism, the new scientific *weltanschauung* was not itself better disposed toward the self. Despite the supposed introspection of the Cartesian *cogito* and its defense of Mind, the legacy of science was to complete what the Greeks and medieval rationalism had begun. The great ingenuity of Cartesian rationalism, and with it that of Spinoza, Newton, Bacon and Enlightenment thinkers as a whole, was to convert the empiricism of Galileo into the *reduction scientiae ad mathematicam* whereby all that is sensuously given was replaced, by Mind, with a system of mathematical equations predicated on a universal *logos*.[17]

Herein too lay the paradox of the new world view. On the one hand, it was predicated on a newly fashioned humanism in which "the ultimate point of reference [was] the pattern of the human mind itself, which assures itself of reality and certainty within a framework of mathematical formulas which are its own product."[18] On the other hand, in order to solve the dilemma of self-doubt that arose from such introspection (i.e., *cogito ergo sum*), the self itself was eliminated by that necessity which knows only the persuasion of mathematical logic. Mind, that is, was not the mind of anyone in particular, but rather a universal mind (the "pattern" of the mind), the pure thought of which is typified by mathematics and geometry. The idealism of the new world was thus never *subjective* idealism. The paradox of that idealism, furthermore, was that in order to defend its own brand of humanism, the new sciences had to rid themselves of any anthropocentric taint.[19] Thought became the center of being, but had nonetheless to rid itself of the familiar enemy: the potential assertion of self, its suspect will, unreliable senses, and fearsome accountability.

Such was the case even as the eighteenth century "mind of the Enlightenment" flirted with the subjective implications of empiricism and psychology.[20] Though advocates of sense perception as the fundament of knowledge, neither the British empiricists nor their French Encyclopedist successors became advocates for the self. In neither case was the individual mind understood as the *source* of knowledge and action, nor was the individual mind responsible for its content. On the contrary, empiricism meant either the "sensationalism" of Locke (predecessor of modern-day behaviorism) wherein the individual mind merely received stimuli from the material environment, or the "idealism" of Berkeley wherein the source of ideas and thoughts in the mind was the "incorporeal Spirit, the Author of nature." Carried over into the psychology of Condillac, moreover, empiricism found fulfillment in the view that, once "analyzed," individual acts of the mind are in no sense original, but wholly derivative.[21] Here as elsewhere, the individual succumbed to the Enlightenment fascination with the *esprit de systeme*.

That spirit or mind, "the rationalistic postulate of unity" became the sine qua non of all Enlightenment thought.[22] Identified by some with geometric thinking, but perhaps best described as a mode of thought which, using the calculus of mathematics, seeks Unity above all else, the chief legacy of the Enlightenment was a *weltanschauung* that reduced all being (man as well as nature) to a common denominator: the method of analysis.[23] In that spirit, nature became subject to *logos*, which is to say, the Mind of generic Man. Yet, in that spirit too, men-in-particular, the concrete individual, also became subject to *logos*. Herein lay the intellectual foundation for Comtean positivism, as well as the meaning of John Stuart Mill's famous proposition that science need not recognise any methodological distinction between man and nature.[24] If the humanism of science here remained committed to the supremacy of Man, it had little tolerance for the idiosyncracy, willfulness, and irrationality of men.[25]

The issue at stake here has usually been termed a problem of "determinism," but its root is the struggle against the self.[26] Its geographical consequence, furthermore, has been a human geography everywhere focussed on generic man, man-in-general, and man-in-mass; a geography of the multitudes rather than of anyone in particular. Not even the bias of "areal differentiation," and "regional geography" was sufficient to overcome that essential methodological inclination, for as Richard Hartshorne noted some eighteen years ago:

> our traditional assumption overlooks the important influence which individual persons may exert on the motivations and actions of hundreds of millions of other people, with resultant consequences of major importance in the geography of areas, small and large. The face of Europe today would certainly be

different if Julius Caesar had not survived his Caesarean birth, of if Martin Luther had been killed by the lightning which struck him at the age of twenty-two. Thousands of lesser figures have influenced small and large communities, and thereby left the mark of their leadership on the geography of every country, even if their names are no longer known.[27]

Though Hartshorne was undeniably correct in this general assessment, the point here is not so much that the traditional assumption "overlooked" the issue. Rather, it submerged the issue by right of Reason and the *esprit de systeme*. Indeed, it eliminated the issue most effectively by the adoption of a social and psychological method which aims to discover the role of decision making processes as they relate to human behavior in the landscape.

In this context, we need not raise the spectacle of location theory, central place paradigms, or the logic of spatial analysis in order to discern the merely vistigial position of the individual and the self in the human landscape. The best examples are those which themselves ostensibly proclaim "humanist" methods and goals. Thus, for example, in the interpretation of urban landscapes David Harvey recently informed us that "the city contains all manner of signals and symbols [and] we can try to understand the meaning people give them."[28] At first glance, this would seem to spell the occasion for an idiosyncratic and even "subjective" look at the various ways different people perceive and mold their environments, thereby rendering them "meaningful." At the least, we might expect an interpretation of the meanings people *give* their landscapes. In fact, however, what we are given instead is a panoply of qualifications. First, the landscape symbols themselves are deemed meaningful only in the light of their most general concatenations. What is more, only the "messages people receive from their constructed environments" acquire significance here, for only they can be measured with any accuracy as regards behavior. As for anyone in particular, any one individual that might have constructed the landscape, Harvey is quick to add that he "doubt[s] very much whether we will ever truly understand the intuitions which lead a creative artist to mold space to convey a message."[29]

The key words in that last sentence are "intuitive" and "understanding," for they underscore the common rationalist *belief* that any individual source of creativity is necessarily inaccessible to the "understanding," hence ineffable (and in its turn, potentially dangerous). That is, "understanding" requires some "very general methodology (in this case) for the measurement of spatial and environmental symbolism."[30] And yet, even here, whether as mere vessel for the reception of messages or as inaccessible vestige, the self remains lurking in the background. Not even that "very general methodology" can do without the very particular individual either as articulator and archetype of spatial meaning, or simply in the person of the methodologist. What is more, the failure to understand the intent and responsibility of the individual in no sense here mitigates the "fact" of "intuitions which lead (some individual) to mold space to convey a message." It merely admits to a failure of method. Indeed, by his very doubt, Harvey proclaims the individual (i.e., the "creative artist") as the source of landscape meaning. That his mode of analysis does not permit an understanding of the role of the individual is not to deny that role, but to deny the method of understanding.

Self and Responsibility: The Ethical Issue

The further tale of the battle against the individual, against subjectivity, and against the self on the part of the sciences is undoubtedly well enough known to obviate the need for further treatment here. Its manifestation in the rise of absolute or objective idealism, materialism, the movement toward objectivity, logical positivism, modern nominalism, and the manifold forms of determinism in the social sciences, history, and geography cannot, in any case, be given adequate study within the limits of this article.[31] For our purposes the salient point remains only that the modern *weltanschauung* is filled with peril for the individual and the self. For that reason alone, if for no other, modern man has become increasingly aware of his own or merely vestigial self. As evidenced by David Harvey's urban landscape, even as the self becomes ineffable, it is nonetheless missed, else why even mention its "intuitive" existence? Indeed, why mention the individual at all?

Whatever "intuitive" reasons Harvey might have had, the essential cause for the missed self is ethical. What is perilous for the individual and the self is the loss of *freedom*. What is at stake here is that freedom upon which the issue of responsibility depends for good as well as evil, for the rational as well as the irrational. "Anxious to avoid all personal (and) above all, all moral judgements," as Isaiah Berlin noted in the case of historians, the modern world view tends "to emphasize the immense predominance of impersonal factors in history, of the physical media in which life is lived, the power of geographical, psychological, social factors which are not, at any rate consciously, man-made, and are often beyond human control."[32] In this spirit, men—concrete individuals—serve a role, but one defined as "the means," the instrument, the manifestation . . . of some vast all-embracing schema of eternal human progress, or of the German Spirit, or of the Proletariat, or of post-Christian civilization, or of Faustian man, or of Manifest Destiny, or of the American Century, or of some other myth, or mystery, or abstraction."[33] What is left of the individual, in short, is a vestige, a necessary medium for the workings of one or another "force" (i.e., reason, irrationality, progress, history, etc.), but having little or no responsibility either for himself, or for the landscape and history which he serves.[34]

Apart from the fact that such notions fly in the face of virtually all our ideas, whether "conservative" or "liberal," about justice and the individual in society, the essential difficulty here is a matter of ethical and logical consistency. That is, unless we attach some fundamental significance to the notion and reality of free, individuated, and even willful subjectivity, we will not be able to distinguish beween necessity and responsibility. Yet, it is this most ancient distinction, summarized here as the distinction between Athens and Jerusalem, which underscores our ability to make judgments and assign values.[35] If the one disappears or becomes merely latent, so goes the other, and with them the very issue of judgment and value becomes moot. The none too flagrant danger here is that in the process we ourselves, as well as all our ideas, beliefs, and feelings (including those of Harvey and Humboldt, Marx and Hegel, and anyone else), whether rational or irrational, lose any discernible value. As we ourselves become just so much grist for the mill of the system, mere media and archetypes of one or another category or process, *our value* becomes ineffable.

This alone should suffice as good cause for something more than the merely tacit recognition of the self in the landscape of man. Unless we are prepared to accept our own irrelevance, we require a logic and a method prepared to assert the decisive role of the individual in the making and meaning of the landscape. Phrased differently, we require a logic that reverses the position of man in the environment so that the latter constitutes the media of the former. Yet, even as the ethical argument commends and even demands that reversal, we are nonetheless faced with a difficult problem. Constrained by an intellectual and an ethical legacy fraught with peril for the self and individual responsibility, any endeavour to identify the role of the individual will be difficult in the extreme. If Athens did indeed overwhelm Jerusalem, so that not only G-d, but also the self and its willful subjectivity and responsibility died in the process, we have been left with landscapes whose authors remain merely latent and ineffable. How then can we hope to identify, let alone understand, the logic and method of a "biography of landscape?"

A Biography of Landscape

We could, no doubt, turn immediately to ideas and methods opposed to the ethic of latent self. The history of the war against the "I" is, after all, almost equally a history of movements poised to defend the self, subjectivity, and the individual. Yet, most of these movements arose as protests against, and therefore confirm the erosion of, the self. Many, too, such as subjective idealism and romanticism, came more as intellectual *culs de sac* than viable alternatives; important more for their critiques than their contributions. Others, like the radical empiricism of William James, Husserl's phenomenology, and modern existentialism appear more fruitful, but they remain on the intellectual frontier awaiting methodological viability by means of substantive contribution.[36] Even as we might derive certain ideas from these movements, in short, we cannot simply turn to them for an answer. Neither can we turn simply to some "other side" in the battle against the "I," for the issue of logical and methodological legitimacy is too complicated for any successful "leap" into the subjective.

Limits of Culpability

However much those inclined to make such a leap might protest, the fact remains that the sciences have successfully shown the limits of individual culpability. We understand now better than ever before that human and individual choice, freedom, will, and responsibility are undeniably constrained. That man is an object in nature, a function of bio-chemical drives, a victim of the DNA molecule, and both a subject and an agent of natural change, cannot be disputed lightly. Neither can we deny that human beings as such and as individuals live out their lives in close accordance with hereditary, physical, psychological, social, educational, and broadly environmental conditioning. Nor is it deniable that virtually all our thoughts, feeling and actions are subject to a mode of classification and analysis that renders ourselves merely latent in the environment. We cannot, for that matter, even deny the ethical merit of the containment of choice and responsibility. Without the "morality of the system," we would still probably punish the insane for their insanity, the poor for their poverty, and the sick for their illnesses. Indeed, without the containment of choice and responsibility, the very ideas of justification by circumstance and rehabilitation would be pointless.

Taken at face value, all such "facts" of limited culpability are encumbrances along the path leading to a biography of landscape. They persistently remind us that the individual is a surrogate for and archetype of one or another set of environmental factors, historical movements, socioeconomic forces, and psychological drives. They remind us too that the landscape is equally a product of such processes. Taken in this light, even the presumed "classic" landscape of individualism, the American frontier, owes its "character and pattern" to processes for which no one in particular can be held accountable. That is, the making of the American landscape derives its impetus and character from a supposedly abundant, "free and open" environment, as well as from the archetypical "frontiersman" and "pioneer."

As Frederick Jackson Turner was to note in the case of the American West, the essence and hero of the frontier was that blend of personalities composed archetypically of Henry Clay and Andrew Jackson.[37] In this sense, the essence of the frontier landscape, no less than the continent as a whole, was the spirit of a burgeoning capitalism mixed with an attendant populism. If, as glorified by Walt Whitman, the landscape of America was made in the wake of an unrelenting individualism, it was here too nonetheless an "individualism" fashioned by "modernity," contrived of the progressive urge to overcome the distances, and driven toward the enlightened elimination of distinctions in the name of unity, progress, and nation.[38] Indeed, the real hero of the landscape was Whitman's "modern man," the Promethean agent and enlightened purveyor of the Spirit of Progress. In this view, nothing so much characterizes the making of the American landscape as the grand engineering feats that served to blend the regions and cross the distances evidenced, for example, in the spectacle of the Brooklyn Bridge and, more dramatically, by the "smoke belching, fire breathing, iron horse" that cut its swath across the continent. As for the western frontier, it too was "won" by "modern man"; by the artifacts of an industrializing society committed to the progressive urge.[39] Socialized by the repeating rifle, sculptured by the deep plow, sealed by barbed wire, or integrated by the railway, the western landscape was virtually engineered by *things*. Or, at least, so it might seem.

For reasons such as these, the issue of landscape authorship—the who behind the facts of geography—becomes nebulous. Swept into the current of social, economic, political, technological, and ideological contexts over which he has little or no control, the individual survives at best as an archetypical figure, important less for who he is and what he does and thinks, then for what he represents. Thus it is that the family Roebling remain, like their bridge over the East River, suspended in the context of an industrializing architecture.[40] Thus it is too that the almost wholly "artificial" landscape of New York City can be "explained" in terms of nodal theory, the great transoceanic migrations, or even the Progressive Movement, not to mention capitalism, class struggle, or simply human avarice. In the light of such contexts, the offspring of those migrations, the merchants, entreprenuers, and victims of the market place pale as meager representatives. They—the Harrimans and Rockefellers, the denizens of Tammany Hall, the brothers Seligmann, Simon Guggenheim, and among the equally famous, Fiorello La Guardia and Robert Moses—figure as ghostly inhabitants, merely latent in contexts for which they are the human archetypes, but for which they cannot be held responsible. In this fashion the landscape itself also appears meaningful only in the context of one or another "tide," "current," "force," or "movement." Here too the landscape, much as does the environment, appears somehow "necessary," somehow "inevitable," somehow "determined," or even "ordained to develop" in a certain direction.[41]

This is not, however, the necessary or lone conclusion to be drawn from the "facts" of limited culpablity. On the contrary, seen first from the vantage of individual responsibility, all such "facts" merely underscore *contexts* for *self-expression*. Viewed in this manner, the great merit of the sciences, and especially the social sciences, is that they have more than ever before clarified the *contextual media* through which and by means of which individuals and specific groups mold their environments to create (what are at least for them) meaningful landscapes. We cannot ignore the "facts" of limited culpability, because it is only through some context—invariably not self-determined, if often or always self-interpreted—that individuals express their particularity and partiality. At the same time, however, we cannot safely ignore such particularity and partiality—the stuff of subjectivity—lest we lose the authors and their meanings in the context.

The Analogy of Authored Landscapes

Rather than belabor an already much overworked argument about the relationship here between subjective and objective modes of human thought and action, perhaps it will suffice to explain that relationship by means of a convenient analogy. Viewed in terms of individual responsibility, the created landscapes of man are much like any other product of human creativity. They have much in common with the manifold forms of human art and artifice. That is, they are constrained by need and context, but they are also expressions of authorship.[42]

On the one hand, the "facts" of limited culpability in the making of landscape are here much like the media of art and literature. Constrained by the media—by paper and canvas, watercolor and oils, pen and brush, soapstone and granite, knife and chisel—artistic creations are always a function of some physical, not to mention aestheic, intellectual, or social context. Similarly, the images, symbols, metaphors, and the very language of literary expression are never wholly subjective and unique, lest they also be totally incomprehensible. In this sense, art and literature, much as the artist and writer, always reflect "situations" of reference, i.e., the physical, social, cultural, historical, and geographical contexts of self-expression and consciousness.[43]

On the other hand, having arisen from some context, art and literature also reflect an "engaged upsurge in a determined point of view."[44] The images, symbols, metaphors, and most of all the meanings, whether visual or literary, are always references to something on the part of *someone*—the author. If that "upsurge" is always "engaged in" some context, the product itself is equally the function of some author's intentions, perspectives, aspirations, inclinations, or broad partialities. In this manner, every work of art imposes an order on reality, and however much that order may be contingent upon its contextual origins, it is nonetheless the responsibility of its author. It is his responsibility, furthermore, because he, first and foremost, gave it meaning. Even if successive generations of critics reinterpret that meaning in the light of their own contexts, the author's responsibility does not change. Neither does *his* meaning change without a change in authorship.

In much the same fashion the created landscapes of men are also contingent upon contexts, but the responsibility of authors. Here, as in art and literature, contexts provide the means, whereas authors provide the meanings. Context and author, environment and individual, object and subject are all, in this sense, correlative, and can be separated only to the detriment of one or the other. Landscapes without contexts would be like books without pages and language. They might exist in the unbound imagination of some author, but they could not be read by anyone, including the author himself. Similarly, landscapes without authors would be like books without writers. They too might exist, but only as bindings filled with empty pages.

Authors in Context

If the question of subjective versus objective modes of thought and action is here rendered moot, this is not to suggest that the relationship between author and context can be put aside. On the contrary, that relationship constitutes one central issue for any biography of landscape, much as it does for biographies in general. Just as we cannot understand Shakespearean drama without some understanding of Shakespeare in the context of Elizabethan England, sixteenth century stage craft, and late medieval English, so too can we not understand the making or the meaning of the English landscape without some understanding of Elizabeth I, the Tudors, the landscape gardeners William Kent and Lancelot Brown, and even our old yeoman friend in the context of preindustrial society and the enclosure acts. That is, no biography of landscape would be feasible, let alone complete, were it to indulge in the purely solipsistic, or ignore the contexts of authorship. At the same time, however, the critical point of any biography of landscape is not to lose the author in the context.

David Harvey and other social scientists to the contrary notwithstanding, the so-called "intuitions" that lead individuals to "mold space to convey a message" are by no means inaccessible or ineffable. On the contrary, they are everywhere evidenced by the way individuals explain, rationalize, or describe their intentions. The datum of those "intuitions" are, for example, accessible in diaries, letters, books, poems, paintings, and in the broad archival collections of individuals. So too are they accessible by means of interview and discussion. That is, we can probe the intentions of individuals, whether rational or irrational, right or wrong, good or bad, to find the meanings they ascribe to a landscape already given, and to find the means whereby they mold their environments to create meaningful landscapes. Rather than depend entirely on the way people behave or respond to stimuli for an understanding of the landscape, we can engage the individual on his own terms and in his own context. We can, in short, discern biographies of landscape most directly by examining what individuals have to say about themselves and their contexts, as well as by examining what others have to say about those individuals.

In this fashion, for example, if we cannot understand the landscape of modern New York City without reference to its changing economic base, the

Authored landscape: The three-pronged Triborough Bridge, most of the expressways, urban renewal housing projects, several of the parks, and practically everything on Randall's and Ward's islands were shaped by Robert Moses. (Caption adapted from Robert Caro; photo from the New York *Daily News*)

rise of an industrial economy and architecture, and the impact of the Progressive Movement, so too will we not understand either the making or the meaning of that landscape without reference to the brothers Roebling, Louis Sullivan (the "father" of the skyscraper), J. P. Morgan, the Rockefellers and Harrimans, and especially Robert Moses. Indeed, as Robert Caro has so amply shown, the landscape of New York City (and much of upstate as well) simply cannot be understood without focusing on the role of Robert Moses and his spectacular empire, the Triborough Authority.[45] "Landscape by Moses," as Caro called it, is New York. The fact of his authorship is overwhelmingly evidenced in the modern history of the city. It is, furthermore, a record of authorship abundantly accessible, as testified by Caro's detailed bibliography and notes. If there is any doubt as to the fact of landscape authorship, one need only read Caro's 1162 page text, all of which testifies to the single-minded drive of Robert Moses to shape and make the landscape of New York in his own image.

Neither is New York alone in this regard. What might the cityscape of modern Shanghai be like without the contributions of the Scots Jardine and Matheson, or the Jewish merchants David Sassoon and sons, and the host of nineteenth century Chinese merchant financiers? Indeed, the city—any city, East or West—is a particularly appropriate place for this form of landscape "analysis." As the contrived depository of civilization, the urban *topos* acts as, among other things, the center of literati and elite culture; the place par excellence of human and individual intention and rationalization articulated and stored not only in infrastructural and architectural form, but also in the collectanea of libraries, museums, galleries, bookstores, universities, newspapers, and the recorded memories of both dead and living inhabitants. Almost seething beneath the "greater story" of urbanization and urbanism, beneath the surface features of urban type or category lie the manifold stories of all those inhabitants whose multifarious intentions, aspiration, inclinations, thoughts, loves, passions, and hates made and continue to make the city. To miss their reality is to miss the "concrete multicolored reality of individual lives" about which Tolstoy so emphatically wrote.[46] To miss them, furthermore, is to miss the *fact* that *they themselves*, not their phantoms, not the "likes of" them, or their archetypes made the landscape.

Understood in these terms, the biography of landscape has as its central concern the role of individuals—authors—in the making of landscape. Its central geographical task is simply to follow through on Hartshorne's complaint that geographers, including those engaged in the study of "decision making processes," have underestimated the importance of key individuals and "thousands of lesser figures (who have) left the mark of their leadership on the geography of every country, even if their names are no longer known." If the rationale and argument for the effort to redress that situation has to do in large measure with ethical and logical appeal, it also has to do with historical "fact." We can argue that World War II "might have occurred" without Hitler, that the War of 1812 "might have happened" without Napoleon, or that New York would have developed regardless of Robert Moses, but the *fact* remains that they did not occur without them. Rather than deal with what "might have been," in short, the biography of landscape deals with what was and is: notably the concrete world of individuals in their contexts, a world of authored landscapes.

Authored landscape: (above) The Shanghai "Bund," 1976 (M. Samuels); (below) David Sassoon and sons; (right) Chinese merchants (Shanghai Chamber of Commerce, 1907)

DAVID SASSOON & CO. LTD.

Authored Landscapes

There are, to be sure, a number of methodological problems associated with the *study* of authored landscapes. Not the least of these arises from the battle between idealist and materialist modes of explanation. Simply phrased, the ancient distinction between a world imagined, and a world lived-in (between the world of mind and the world of reality) persists to condition the way in which we come to understand the nature of authored landscapes. If the logic of landscape biography renders this distinction moot, the method of that biography must account for the relationships among these two worlds. Viewed in terms of authorship, the world of mind and the world of reality are *equally* subjective and objective, and it is the special task of the biographical method to expose their mutual interdependence.

Once again, rather than indulge the heady debates between idealists and materialists, it will perhaps serve our purposes as well to employ an analogy with art and literature. That is, the distinction most appropriate to the study of landscape biography is one between landscape *impressions* and landscape *expressions*. Neither of these are simply subjective or objective.

Both require an author in context. They are, however, different products of authorship.

Landscapes of Impression

Landscape impressions, by definition, belong to and arise from the thoughts of someone. We encounter here all those poets, artists, philosophers, theologians, thinkers, and others (including the rest of us) who, in perceiving an environment already given, reshape it into an image, give it some special meaning and content, and not only transform "reality" into an impression, but also make it impressive. Robert Frost's New England, Gauguin's Tahiti, Hemingway's Spain, the Holyland of the Prophets, and everyman's home are, in this sense, impressionistic. They are imaginary places in the original sense of the verb "to image." They are also often metaphorical, having to do with hidden meanings associated with or evoked by a place. Sometimes, too, as in the case of Tolkien's Middle Earth, they are allegorical, having little to do with the immediate world of everyday places. Like J. K. Wright's "geographical ideas, true and false," they also have to do "in large measure with subjective conceptions."[47]

Being subjective, landscape impressions are themselves more intellectual than "real," more *about* than *in* the landscape, and are thoroughly biased by their authors. For this reason, the woods of landscape imagery are almost everywhere filled with misbegotten caricatures, all of which testify to the highly aimed nature and content of impressionism. We need not search far to discover the wild record of such caricature. Virtually any nineteenth century geography textbook will suffice to give examples enough. Lest we feel better than our predecessors, almost any discussion of the so-called Third World will also reveal a wealth of landscape caricature. The idea that Asia and Africa had to be "discovered," and once discovered, "opened up to civilization" reveals a bias built on landscape ignorance. Equally, the notion that the landscapes of China or India are "underdeveloped" betrays an image built on cultural biases that ignore the fact that, if anything, the Chinese and Indian landscapes are "overdeveloped." The standard notion of a "developed landscape," is little more than a euphemism for our own, Western landscape.[48]

Frequently, no doubt, landscape caricatures are dangerous. Northern European images of tropical fecundity, for example, often found reflection in notions that tropical environments breed indolence, that tropical peoples do not value life as much as temperate zone peoples, or that the tropics require leadership by and tutorials in "advanced" (read, "temperate") civilization. Such notions were, of course, commonplace justifications for colonialism and imperialism. Similarly, exaggerated images of one's own place, or the importance of one's own landscape frequently involves the exclusion of and attack upon others who figure as alien elements within or outside that landscape. Indeed, exaggerated ideas about the importance of roots in a place, the mystique of landscape attachments, have their counterpart in caricatures of uprootedness. As in the famous myth of the Wandering Jew, landscape caricatures can by used to exploit and persecute others, as well as defend and circumscribe one's own meaningful places.[49]

The best examples of landscape caricature today are undoubtedly found in the literature of nationalism. The modern history of nation-states is filled with exaggerated images about homelands, motherlands, and fatherlands. They are exaggerated both by means of poetic license applied to the "we" of some identity with place, and by exaggerating the conditions of an alien "they." When, for example, in the case of Margaret Atwood's novel *Surfacing*, Canada figures as a virginal, untrammeled, feminine spirit of the forest, the landscape image can be said to have its intellectual roots in topophilic myth and in the perceived reality of Canada's vast, unsettled interior.[50] That she should invest her home, Canada, with mythopoeic affectation is a reflection of her affections, and a matter of metaphorical or poetic license. That the image is a caricature of Canada, furthermore, is only to say that her Canada exists more in the imagination than in "reality," or that it ignores a Canada touched by the CBC, chainsaws, skidoos, canned soup, and cities. It is a landscape of lost, but aspired to, innocence; a modern day variant of romanticism. But it is also more than this. Threatened by its alien counterpart, an American demon archetype of technology, the context of the image—the rape of the forest—is but a scarcely disguised form of chauvinism. Not technology per se, but *American* technology becomes the behemoth rapist; and it is not technology or the machine, but the United States which is aimed at overwhelming the innocence of the forest. If the image of Canada is here benign, its intent is political. As topophilic myth and political message, its lineage can be traced to Wagnerian hyperbole and Fascism.

The evidence of hyperbole, exaggeration, or caricature in landscape imagery, however, does not mean that landscape impressions are lacking in an objective content. On the contrary, if subjective in origin, landscape impressions acquire an objective content insofar as they have a history: a history of authorship, diffusion, and impact. That is, whether benign or odious, landscape images have an objective content as: (1) they can be attributed to someone who (2) created, obtained, or conveyed an image in a context which (3) is shared with others. The image can be explained in terms of an historic someone somewhere. It can be explained in terms of a reality charged and often supercharged with attachments to and identities with places, people, and things on the part of some author in some context. In the process, the image itself acquires an objective content, because it too has a history.[51]

Identified with and explained by the concrete circumstances of their authors and contexts, landscape impressions here have at least an intellectual history. In principle, that history can be traced, and it is one central goal of a biography of landscape to accomplish that task. Admittedly this is not always an easy task. Especially as we seek out the original authors of impressions, we are often left instead with the authors of contemporary usages, the lineage of which has been reduced by time to little more than a context. Like the words employed to describe or explain a landscape, the original authors, contexts, and meanings of landscape images are often lost, but residual as contexts for the making of some ersatz or otherwise loaded image. The image may become, as it were, part of the media for the making of its likeness in the impressions of others. Here too, however, the image acquires an objective content as it is shared, promulgated, and changed by other authors to suit their own purposes. The image may become part of the media for the making of shared landscapes.

If the biography of landscape impressions has to do in large measure with a type of intellectual history of environmental or spatial ideas and images, it is also implicitly an exposition on the historic origins of behavior—individual and social—in the landscape. More often than not, the intent behind any study of ideas is to explain something about their impact on the lives of those who share the ideas, or on the social landscape as a whole. Landscape impressions here become the contexts for the making of landscape. At this point, we begin to move away from an explanation of imagery, and towards an explanation of the landscape itself. We move, in short, toward the landscape of expression.

Landscapes of Expression

Like the case of landscape impressions, we are faced here with a number of problems related to the discovery and meaning of authorship.

That is, landscape expressions are, by definition, expressions of something on the part of someone. Here, however, the facts of authorship are made more complex, on the one hand, by virtue of the frequency with which social landscapes are the products of pluralities, rather than particular individuals. On the other hand, even as we come to identify particular authors, there remains the problem of how images, whether of the landscape or of something else, become the contexts for the making of landscapes. That is, all impressions do not necessarily find reflection in the landscape, anymore than all ideas or wishes are realized. The world is, after all, filled with those who dream a dream of places and utopian landscapes, but never do anything to fulfill those dreams. For this reason too poets, artists, philosophers and others are often poor sources for the explanation of an extant social landscape. Though they may explain the meanings of landscapes, they do not necessarily tell us anything about the making of landscapes.[52] In both cases, an understanding of landscape expression requires a method of identification; a means whereby we identify authors *in* the landscape.[53] It requires too a method of explanation; a means whereby we explain the process in which ideas become contexts for the making of landscape.

IDEAS IN THE LANDSCAPE   Whether by virtue of the influence of some author, or by being impressive, ideas and images are often themselves nascent in the environment as contexts for decision making, and for the making of social landscapes. How many of us, for example, have wished or acted on the desire to wander off to Gauguin's Tahiti, or to the Holyland of the Prophets only because of their impressive imagery? What would be the fate of the contemporary jumbo-jet tourist economy were it not for such impressive imagery? When in the 1850s thousands paid the handsome price of 25¢ to see such ridiculously romantic impressions of the American frontier as the Dickensen-Egan "Panorama of the Grandeaur of the Mississippi Valley," how many decided there and then to make their lives in the frontier? What would have been the history of the "Jeffersonian Ideal" and the American landscape had not Ben Franklin and others been inspired by Quesnay's Physiocratic exaggerations of the ideal Confucian landscape?[54] Where, for that matter, would the modern landscape of North America be today had it not been for the utopian imagery of the Progressives? The German fathers of that doctrine, Karl Haushofer and others, however, were not alone in this regard.[57] Much the same sense of culture-in-place with equally chauvinistic connotations can be traced to the history of French nationalism.[58] The idea and its consequences are familiar too in the history of "Mother Russia."

Similarly, the American principle of isolation, its attendant geographical rationale, and its manifestation in particular doctrines (*e.g.*, the Monroe Doctrine, Manifest Destiny) can also be cited as a case of landscape image turned to political and economic cause. Here too landscape imagery became geopolitical doctrine with various, and often devastating consequences in the landscape.[59]

Parochialisms of all sorts, no doubt, acquire political, economic, and legal sanction to influence the making of landscapes. The ideology of place, for example, finds communal fulfillment in nationalism and in various scales of parochial attachments, but it also finds social and individual manifestation in the doctrine and value of private property. That is, the doctrine of private property has, at minimum, its origins in attachments to and identities with particular places as one's own. Home, estate, land, and the things within one's zone, as it were, acquire legal, economic, and political sanction to become a doctrine of privatism. The origins of the doctrine can furthermore be traced through the history of law, especially American, British, French and eventually Roman Law. The impact of the doctrine can also be traced in the form and pattern of land use, as well as the development of particular institutions on the land made to protect, facilitate, or sustain the doctrine (e.g., banks, real estate companies, mortgage houses, insurance companies).[60]

There are, no doubt, other forms and means whereby landscape images acquire ideological content and meaning. For our present purposes, however, we need only note that, whatever the specific nature of political, economic, and legal sanction, once filled with ideological content they constitute the intellectual grounds for action on the part of individuals and groups.[61] They become the intellectual contexts for the making of landscape. Then, and only then, can we speak of landscape expression, for only then can we trace an extant landscape back to its roots in someone's view of place.

AUTHORS IN LANDSCAPES   Ideologies, like ideas, also have authors. The history of their authorship can also be traced. But, as we encounter the expression of ideology *in* landscape, we are faced with a number of problems in the discovery of authorship. Simply put, the history of authorship here is often submerged in the mass social pluralities that are said to be responsible for the landscape. The issue here is less a matter of substance or principle than of method. That is, landscape expressions, no less than impressions, belong to someone in some context. The fact of social plurality does not mean that shared landscapes have no authors. Neither does it mean that no one in particular can be held accountable for that landscape. Rather, what it means is that, in order to discover the history of authorship we need a method or a series of approaches that focuses upon the role of authors in the landscapes, in general and in particular.

The evidence for authorship in general is, no doubt, most abundant in terms of the *design* of landscape. So long as one speaks of design in the landscape, the presumption is that there is or was a designer(s). Classic evidence arises from the examples of gardened landscapes. Gardens are the most evident cases of designed landscapes. For this reason, when Hoskins described the making of the English landscape, he included the role of Lancelot Brown and others in the making of the English garden. For much the same reason, no study of the Chinese landscape would be complete without reference to the Chinese garden and its authors. The latter include not only famous landscape architects of Chinese origin, but also, as in the case of the grounds of the Summer Palace, such well known Westerners as the Jesuit Father, J. Castiglione (Chinese name: Lang Shih-ning).[62] Similarly, no study of the Japanese landscape would be complete without reference to its famous gardens, all of which can be traced to particular authors, and none of which, despite the apparent conventions of design, are quite like any other. In this sense, gardened landscapes are most significant because they clearly reveal their authorship.

Gardens are, however, obviously contrived and essentially idiosyncratic places. Few of us live in gardens, or contribute to their design. In that regard they are not social landscapes, even if they are used by society. Moreover, gardens are not the most appropriate model because they are essentially and purposefully benign places. They are places specifically carved *out of* the landscape as preserves away from society. Were it possible to rid the term "garden" of its benign connotations, however, the principle and practice of design here would have application elsewhere in the human landscapes.[63] Though they are not necessarily "planned" or "rational" landscapes, cities also entail design and have authors. Whether benign or evil, the city is also a "garden," which differs from other gardens primarily in that its authors are plural and involve all those who live in and contribute to the design of the landscape. In addition, cities and the social landscape as a whole are different in that they bear the evidence of design and authorship not only in form, but also in name.

Designations in, as much as designs of, the landscape are also important evidence on behalf of authorship. The designation of area, as in the selection of "choice" areas (the term itself reflects a fact of authorship), are both signals and symbols in the landscape of human intent. Like designs, they also have a history that can be traced to someone somewhere. Thus, for example, as David Ward recently noted, the term "slum" has a history that can be traced to particular people and groups.[64] Similarly, in the history of the selection of "choice" areas, "view" had less to do with the making of such zones than "who."

In this regard, the makers of landscape imagery in the modern context are also often the makers of the landscape itself. Whether as purveyors or makers of landscape designation, real estate agents, brokers, and developers are often good sources for a biography of landscape designation. Just as they convey and perpetuate landscape intentions cast by others, so too do they create, manipulate, and designate the forms and meanings of places. Their hyperbole, as well as the shapes they give to places, literally mold places to convey a message which, presumably, they think is effective. That their designs and designations are reinforced by others goes almost without saying. Indeed, between real estate broker, developer, banker, insurance agent, and client often lies but a thin red line: the line of assigned or designated landscape value.

If the evidence of authorship in the landscape is made clear by the facts of design and designation, we are still left with the problem of identifying particular authors. We may understand that some author or authors were responsible for an event or fact of landscape, but find the task of specific attribution difficult. At this juncture the biography of landscape becomes akin to the task of investigative reportage. It becomes, as it were, a search for those who have obviously, intentionally, or inadvertently left their signatures somewhere.

One place to begin that search is with the names people give their places. Among the ten different modes of place designation described by G. Stewart, one bears directly on the identification of particular authors.[65] Frequently people give their own names to places, or have their names assigned to the landscape by others. The record of landscape authorship is often vestigial, but evident in the veritable panoply of personal names assigned to towns, villages, streets, and neighborhoods, in the landscape. By tracing those names back to their authors, we will often, though not always, find the origins of places. The city of Seattle, for example, is named after Chief Sealth. The name reveals something about the origins of the city, if only that it had some connection with the relationship between Sealth and the white settlers on the shores of Puget Sound.

If place designations are not always useful to the discovery of authors, there are other means to that end. Once again beginning with the extant landscape itself, there is at least one other way in which the names of people are placed *in* the landscape, notably in cemeteries. That is, local cemeteries are often useful firsthand guides to who made the landscape at particular times. A close examination of grave markers and epigraphy is especially useful where other records are incomplete or unavailable. One can, for example, trace long-decayed trading networks in Southeast Asia by following the grave markers and studying the epigraphy of otherwise forgotten Chinese, Arab, Persian, and Western merchants and seamen. Similarly, one can walk through the English, Persian, Jewish, and Chinese Christian cemeteries of places like Hong Kong and discover the famous, infamous, and obscure members of the colonial and resident establishment, carved in stone. Here too the evidence of the family plot can be most revealing of the many small and large dynasties that made the landscape. One can, if one wishes, begin to trace the *persona* back to their origins, and to the impact they had on the landscape.

There are, of course, other more textual sources for the discovery of landscape authors. That is, as someone filled the landscape with designs—laid out the street pattern, railway lines, water lines—and determined where and what would be in the landscape, so too is the record of their intentions often available to those who care to look. City and county archives are almost everywhere in North America filled with the datum of authorship. It is available in surveyor notes, city and county council minutes, newspaper editorials and columns, corporate stock holder meetings, promotional literatures, and in various other public and not-so-public archives. Here, as elsewhere, those who made the landscape have left a record of their ideas and intentions about where things ought to be.

Other textual sources for a biography of landscape can be found in city directories, social registers, the society and personal columns of local newspapers, and even in telephone directories. Indeed, in cases outside the modern Western world, telephone directories are often very useful guides to local decision makers. For that matter, they are often such good guides to those who make decisions in and for the landscape that in many places, as in the Soviet Union and the People's Republic of China, they are unavailable to any but those with security clearances. In Western societies, a similar function may be performed by professional directories, specialized newsletters, and the membership lists of exclusive clubs.

The use of these and other similar devices, however, brings us to one critical issue in the biography of landscape. On the one hand, the nature of much of this material is aimed toward those whose names have been deemed worthy of record. They are not egalitarian but elitist sources. They often fail to include those who did not own property, were not socially prominent, or simply did not have telephones. They miss transients, new-comers or strangers, criminals, the "outsider," and of course, they neglect the "poor." On the other hand, even with this built-in bias, the panoply of recorded names suggests that we still require some additional means to disaggregate particulars from the mass. If only to focus on those whose impact has been "greatest," "most evident," or more than "commonplace," a biography of landscape almost necessarily indulges an elitist sentiment.

Individual and Elite in the Making of Landscape

Like most terms of social designation, the term "elite" is loaded. In ordinary usage it presumes a measure of prominence and a social hierarchy in which elites lead at the heights, while others remain dormant or inactive in the valleys. That is, certain individuals and groups of individuals marked by wealth, education, or some criterion of social position and status invaria-bly stand out from the crowd as primary innovators or purveyors of choice and decision in society. They may be village or small town, national or international elites. They may be monopolistic or pluralistic, centrist or democratic, permanent or temporary, economic, intellectual, artistic, reli-gious, or political, and they may operate on the basis of consensus or com-petition.[66] Whatever their scale or nature, the ordinary meaning of the term "elite" refers to those who occupy positions of authority in the sense that others follow, listen to, and are influenced by their choices and decisions.

In this sense, many of the examples already employed in this paper are elitist in nature. Many authors of landscape are elites of this sort. But this is neither necessary in principle, nor ordained by the method of a biography of landscape. On the contrary, as in its original meaning, the term "elite" (Lat: *eligere*, Old French: *eslire*) refers generically to the ability "to choose." *Choice* is the central criterion of elitism. To be sure, choice, as in the selection of residential areas, is undeniably often the perogative of the few. Most of us are given little choice in the matter of where we live or how we live. But, as discussed earlier, the ability to choose, whether imple-mented or not, is the essential premise of the idea and reality of human and individual responsibility. That premise, furthermore, constitutes the sine qua non of a biography of landscape, for such a biography demands a view of man and of men predicated on their responsibility for the landscape.

The ability to choose is true for "elites" at the top of the social hierarchy, but it is also true for *the elite*, i.e., generic man and men-in-particular, wherever they stand on the social ladder. Every individual here constitutes an elite who molds and gives meaning to an environment by virtue of choice. To be sure, the implementation of choice is often limited to those who, for one reason or another, have the means to overcome or escape social, economic, political, legal, educational, physical, or other constraints. Indirectly, the geography of socioeconomic indicators is undeniably a geography of those with greater or lesser means to effect choice. But even in the most socially limiting circumstances of birth, race, wealth, education, or position, individuals ceaselessly emerge to mold and create their own landscapes. The sign posts of individual expression are hardly hidden for those who care to look. They are there in the literature, art, dance, and theater of the "folk," the poor, and the "masses." They are evident in the vernacular poetry, graffiti, and media of the ghetto. They are there in the dramatic and not-so-dramatic *persona* of the Lower East Side, Harlem, and Watts. They are there too in village elders, clan spokesmen, and in story tellers. They are, in short, almost everywhere. Indeed, no biography of landscape would be complete without them, not because they constitute the "majority," or because we should be "egalitarian," but because they provide such ample testimony on behalf of authorship despite and even because of their limited responsibility.

Taken to one logical conclusion, the evidence of individual authorship amidst the almost total annihilation of choice and responsibility is the most powerful testimony we have to the fact of authorship. Even in the most restricted environments, as in the Nazi death camps, individual choice is not absent. On the contrary, both on the part of prisoner and captor, there are hierarchies of responsibility.[67] Here too the ability to choose; to choose to live or die, kill or be killed, torture or not to torture, and to create landscapes that either shut out or hem in the reality of concentrated inhumanity is evident. This is not to say, of course, that everyone in concentration camps could choose. Rather it is to say only that those who had no choice, those who were stripped of their responsibility, and those who were overwhelmed by their environment were also, by definition and in the most brutal sense, *dehumanized*. If the victims became mere objects, perhaps their Fascist torturers became automatons. But, then too, automatons are also objects lacking in choice and responsibility. If neither victim nor perpetrator were responsible, if both were dehumanized, just who and what was responsible for the Holocaust? For that matter, just who and what is ever responsible for dehumanized landscapes?

In this last example, the ethic and logic of a biography of landscape is perhaps most clearly identified. Negative examples—the reminders of the Holocaust, antebellum Slavery, Russian pogroms, and such variations on the theme of dehumanization as the Sand Creek Massacre, the slaughter at Hay Market Square, Gulag Archipelago, and more recently the American "adventure" in Vietnam—are unpleasant. In the words of the new American president, they are the events we prefer to "put behind us." But they have the merit of directing our attentions most forcefully to the issue of authorship. They have the merit of proving that, if no one is ever responsible, if we are all but victims of G-d, History, Nature, Reason, or "the System," then we and our landscapes are by definition, lacking in human content. In that event, "explanation" may supercede "attribution," Athens may overwhelm Jerusalem, and all the Mr. Hitlers, as well as all the Mr. Flowers, may become the mere function of some necessity; but we are left too with a world no longer anthropocentric, no longer made in the image of man. If that is the history of the war against the self, then perhaps all that can be added here is that, in principle and in method, the biography of landscape aims in a different direction. It aims in the direction of the anthropocentric, because it is aimed at the self, the individual, the ego, and ultimately toward anyone in particular, including our old friend, Mr. Flower.

Conclusion

There are undeniably many questions left unanswered here. The "theory," let alone the "method," of a biography of landscape is not yet complete, if only because we have here tapped the surface of human choice and responsibility in the making of landscapes. That this biography of landscape is not new, furthermore, is amply testified to by the record of historians like Isaiah Berlin who have more fully traced the roots and consequences of the issue of authorship in history. Indeed, that the method of a biography of landscape derives from historiographic sources has been left largely implicit in this article. If only to outline the theme and issue of that biography, the method, as well as the geographical consequences of authorship, have not here been thoroughly explored.

Those geographical consequences, no doubt, serve as the substance of the issue of authorship in the landscape, especially for those of us who are geographers. Those consequences are manifold, having expression in the matter of landscape design and designation, form and meaning. We can, in this regard, note that the biography of landscape leads to the further elucidation of themes explored and made relevant to modern geography by J. K. Wright, David Lowenthal, Yi-fu Tuan, Annette Buttimer, Clarence Glacken, Paul Wheatley, Donald Meinig, and many others. The inclination behind the biography of landscape, in short, has a geographical lineage, and is in no sense alien to or apart from the intent or purpose of geographic explanation writ large.

If the biography of landscape is different from these and other geographical traditions, it is different primarily in that it focuses on the role of individuals in the making of landscape. It is different too in that it requires greater particularity in the definition of geographical problems: particularly as to *persona* in the landscape, and the particularity of the landscape itself. Like any art or science, a biography of landscape cannot avoid generalizations, but it can avoid levels of generalization that deny the humanity of its subject matter.

It is also true that a biography of landscape is not everywhere or always feasible. Countless millions of faceless and nameless peasants and townsfolk the world over have through the centuries molded, designed and designated their environments in the form of land use patterns and monuments to state grandeur without having left so much as an "X" for a signature. Having disappeared into the landscape much as their corporeal selves have turned to dust in some forgotten stretch of a Great Wall or ancient fendike, we can barely hope to ever recover the datum of their *persona*. Like archaeologists left with only the artifacts of man, social scientists and historians alike are often limited to speculation, inference, and an intuitive feel about the making and meaning of human landscapes. Doomed otherwise to oblivion and a lack of understanding, this is not a dehumanization of the faceless and nameless, but an effort to at least recover the contexts, and thereby define the limits, of their worlds. If we cannot know who they were, we can at least make known what they were not. Perhaps, in those cases, that is all that we can ever hope to accomplish.

If, in this sense, the biography of landscape is limited in its applicability; however, it is limited only by the lack of concrete data in all cases, and by the willingness to search diligently enough.[68] This should hardly bother those of us who profess loyalty to science or to the desire to explain the landscapes of man. What else, other than the data and a willingness to search out that data, should limit understanding?

Notes

1. W.G. Hoskins, *The Making of the English Landscape* (Harmondsworth, Middlesex: Penguin Books, 1970; orig. pub. 1955), p. 98.
2. Ibid., p. 14.
3. Ibid., p. 174.
4. Hannah Arendt, *Between Past and Future* (New York: Viking Books, 1961), p. 64.
5. The argument presented here is taken in large measure from Lev Shestov, *Potestas Clavium*, trans. B. Martin (Chicago: Henry Regnery Co., 1970; orig. pub. 1923), and idem., *Athens and Jerusalem*, trans. B. Martin (New York: Simon Schuster, 1968; orig. pub. 1937). One among the modern pantheon of existentialists, Shestov is perhaps best regarded as the logical better-half of Kierkegaard. As Albert Camus noted, Shestov's work is "throughout wonderfully monotonous, constantly straining toward the same truths, tirelessly demonstrating that the tightest system, the most universal rationalism always stumbles on the irrational of human thought." [A. Camus, *The Myth of Sisyphus* (New York: Viking Books, 1955), p. 19]. Most of that "tireless" demonstration cannot and need not be replicated here. Neither need we concern ourselves here with the ultimate, theological implications of Shestov's argument. Rather, as his discussion of the logical and historical case for the issue of responsibility is one of the most thoroughgoing with which I am familiar, I have sought merely to restate that part of his thesis. A useful companion source is H. Arendt's essay "The Concept of History," *Between Past and Future*, pp. 41–90. For an intriguing discussion of the logical dilemma in contemporary Chinese thought, see Frederick Wakeman, *History and Will: Philosophical Perspectives of Mao Tse-tung's Thought* (Berkeley: University of California Press, 1973).
6. Shestov, *Potestas Clavium*, pp. 152–53.
7. Kenneth Clark, *Landscape into Art* (Boston: Beacon Press, 1961; orig. pub. 1949), pp. 2–4. Shestov, *Athens and Jerusalem* pp. 281, 298–99.
8. St. Augustine, *Confessions* II, 3. Quoted in Clarence Glacken, *Traces On the Rhodian Shore* (Berkeley: University of California Press, 1967), p. 202. See Note 15.
9. This does not mean that there were no countervailing forces. Even in the classical homeland of Reason there were always those subversive few who, like the Epicureans, waged battle against the teleological, ordered universe of the rationalists. A few remained to see the world and the universe as infinite plurality, and to stress the importance of the world of the senses. As in the famous case of Lucretius, there were those who "were never happy without visible demonstration . . . from the world of the senses." Glacken, *Traces on the Rhodian Shore*, p. 67.
10. Clark, *Landscape into Art*, pp. 6–7.
11. Glacken, *Traces on the Rhodian Shore*, pp. 213–18.
12. Clark, *Landscape into Art*, p. 7.
13. Ibid.
14. Glacken, *Traces on the Rhodian Shore*, pp. 302–11.
15. St. Thomas Aquinas, *Summa Theologica* I, Q. 25.2. Quoted in Shestov, *Athens and Jerusalem*, p. 301. The issue of medieval rationalism made of the works of Duns

Scotus, St. Thomas, and St. Augustine is obviously beyond the focus of this paper. However, one point often raised requires brief mention, namely the Augustinian doctrine of "free will." The latter constituted a logical response to the key issue of medieval theology, i.e., the Manichean heresy, or how to account for sin in a world created by an omnipotent godhead. The resolution was the bifurcation of man into rational and willfull capacity; the one accounting for the 'good," the other for "sin." For this reason, classical *gnosis* (knowledge) did not take the place of faith, but became the end of faith. For this reason too, the will of the self became the ground of sin, whereas Reason became the will of G-d, and in turn, the *gnosis* of man. Here, before the rise of Cartesian humanism, "thought" in the form of *gnosis* lost its original sinfulness, and "original sin" was reinterpreted to mean "lust." As Augustine and Duns Scotus noted, "the first sin of man was an immoderate love of union with his wife." Augustinian "free will," in short, did not represent a defence of human caprice, subjectivity, the self, or the will of the individual. It represented rather an attack on the latter, as evidence of sinfulness.

16. Hannah Arendt, *The Human Condition* (Chicago: University of Chicago Press, 1958), pp. 257–60; A.D. White, *The History of the Warfare of Science with Theology* vol. 1, (New York: Dover Publications, 1960; orig. pub. 1896), pp. 130–70, and R. Hooykaas, *Religion and the Rise of Modern Science* (Grand Rapids, Michigan: W.B. Eerdmans, 1972), pp. 29–53.

17. Arendt, *The Human Condition*, pp. 264–73.

18. Ibid., pp. 284.

19. John Passmore, *A Hundred Years of Philosophy* (Harmondsworth: Penguin Books, 1968), pp. 174–200, 320–42.

20. Ernst Cassirer, *The Philosophy of the Enlightenment* (Princeton: N.J.: Princeton University Press, 1951).

21. Ibid., p. 18.

22. Ibid., pp. 22–27.

23. Ibid., pp. 8, 3–36.

24. The methodological issue in the sciences is discussed in detail by Ernst Nagel, *The Structure of Science* (New York: Columbia University Press, 1961), pp. 454–546, and David Harvey, *Explanation in Geography* (London: Edward Arnold, 1969), pp. 44–61. Also see Passmore, *A Hundred Years of Philosophy*, pp. 13–34.

25. The exception was "subjective idealism" based on the empiricism of Bishop Berkeley. Though the latter found reflection in a stronger inductive logic, as well as in an *a posteriori/* empirical view of space (i.e., in Leibnitz), Berkeley had no immediate successors. One can, however, trace the impact of subjective idealism on the sciences in terms of the radical empiricism of William James, the history of Anglo-American pragmatism, and in modern psychology. See, for example, Cassirer, *The Philosophy of the Enlightenment*, pp. 93–133; Passmore, *A Hundred Years of Philosophy*, pp. 95–119; and John Wild, *Existence and the World of Freedom* (Englewood Cliffs, N.J.: Prentice-Hall, 1963), pp. 19–38. For a brief introduction to James see P.K. Dooley, *Pragmatism as Humanism: The Philosophy of William James* (Totowa, N.J.: Littlefield, Adams, 1975).

26. For this reason, the 18th century ideal of human freedom and the beginnings of

social science must be understood in terms of the idea of natural law and the social contract. See Cassirer, *The Philosophy of the Enlightenment* pp. 234–74.

27. Richard Hartshorne, *Perspectives on the Nature of Geography* (Chicago: Rand McNally, AAG Monograph, 1959), pp. 152–53. Hartshorne's own *The Nature of Geography* (1939) was not better in this regard. The reason for such neglect on the part of geographers is that Geography is itself a product of Enlightenment rationalism. As the *esprit de systeme* diffused to and became epitomized by the person of Alexander von Humboldt, it revealed a landscape devoid not of man-in-general or generic *homo sapien*, but of men-in-particular. It is not man, but that spirit—the Spirit of Reason—that rises through the pages of the *Cosmos* to mediate both man and nature. If Humboldt, like his contemporaries, was here a enemy of reaction, a friend of liberty, neither he nor his associates (e.g., the mathmatician Gauss, the politician Thomas Jefferson, and others) advocated the freedom of the self. Despite and perhaps because of his political commitments, as well as his own emotional problems, the *esprit de systeme* became a refuge for Humboldt. For an intriguing biographical discussion of the origins of geography in the person of Humboldt see Douglas Botting, *Humboldt and the Cosmos* (New York: Harper & Row, 1975).

28. David Harvey, *Social Justice and the City* (London: Edward Arnold, 1973), p. 32.

29. Ibid.

30. Ibid. I do not intend here to dispute the value either of behavioral methods in general, or the specific contents of Harvey's argument. Much of the latter, especially as it encourages the detailed study of spatial symbolisms and the role of ideas in the making of landscape, fully accords with the intent of a "biography of landscapes." But, to the extent that those methods and Harvey's argument necessarily operate at the aggregate level only, they also work to the detriment of those individuals who not only make and live in the human landscape, but also author and become the victims of "social justice." Harvey is, of course, not unaware of this issue. His own "philosophical framework" is one "in which space only takes on meaning in terms of 'significant relationships,' and a significant relationship cannot be determined independent of the cognitive state of some individual and the context in which that individual finds himself. Social space, therefore, is made up of a complex of individual feelings and images about and reactions toward the spatial symbolism which surround that individual." At the same time, however, he also proclaims that this itself "would make for a depressing picture from an analytic point of view, if it were not for the fact that groups of people appear to identify substantially similar images with respect to the space that surrounds them and also appear to develop similar ways of judging significance and behaving in space." (Harvey, *Social Justice and the City*, p. 34). The essential difference between Harvey and myself is summed up in his view of that "depressing picture." From the perspective of this paper, group consensus, as such, makes little or no sense unless and until its individual components are determined.

31. Of the many sources most useful to this discussion see especially Passmore, *A Hundred Years of Philosophy*, pp. 174–200, 320–42, and 367–93.

32. Isaiah Berlin, "Historical Inevitability," *The Philosophy of History in Our Time*, ed. H. Myeroff (Garden City, New York: Doubleday, 1959), p. 255.

33. Ibid., p. 257.

34. According to one *au courant* persuasion, this is precisely the historic effect of urban-technological society, i.e., alienation by machine and modern organization. The chief exponent of this view is perhaps J. Ellul, *The Technological Society* (New York: Vintage, 1964). Its more commonplace expression can be seen in such cases as J.K. Galbraith's assertion that "the imperatives of technology and organization, not the images of ideology, are what determine the shape of economic society." J.K. Galbraith, *The New Industrial State* (New York: Signet Books, 1967), p. 19. As Galbraith and others take great pains to show, it is not the *persona* of the "system," not the choices of industrialist, entrepreneur, marketing and advertising executive, or economics professor, and not their values, beliefs, or images, but the "system" itself that determines the structures of modern society. Whatever the merits of this view, it is at least made suspect by the lack of consistency with which it is held by those like Galbraith. It was not that view, for example, which characterized Galbraith's arguments against Richard Nixon during the Watergate episode.

35. Berlin, "Historical Inevitability," p. 252.

36. Though radical empiricism and its links to pragmatism has been left largely ignored, discussion of phenomenology and existentialism in geography has ranged widely in recent years. For a recent critical review see J.N. Entrikin, "Contemporary Humanism in Geography," *Annals*, Association of American Geographers, 66 (1976):615–32. Entrikin's conclusions as to the utility of phenomenology and existentialism are, however, not acceptable to this writer, especially as he ignores the historiographic and highly empirical tradition identified with these two schools of thought. For a different view of the subject see, among others, Marwyn S. Samuels, "Science and Geography: An Existential Critique," (Ph.D. diss., University of Washington, 1971); David Ley and Marwyn S. Samuels, *Humanistic Geography* (Chicago: Maaroufa, 1978); A. Buttimer, "Grasping the Dynamism of the Lifeworld," *Annals*, Association of American Geographers, 66 (1976):227–92; E.C. Relf, "An Inquiry into the Relations between Phenomenology and Geography," *The Canadian Geographer* 14 (1970):193–201.

37. Richard Hofstadter, *The Progressive Historians* (New York: Vintage Books, 1970), p. 95.

38. Walt Whitman, "One's Self I Sing," and "Leaves of Grass." See *The Literature of the United States* Vol. 2, ed. W. Blair, T. Hornberger, and R. Sterwart, (Chicago: Scott, Foresman, 1961), pp. 134–211.

39. Leo Marx, *The Machine in the Garden, Technology and the Pastoral Ideal in America* (New York: Oxford University Press, 1964).

40. Lewis Mumford, *The Brown Decades, A Study of the Arts in America 1865–1895* (New York: Dover Publications, 1971; orig. pub. 1931), pp. 44–48.

41. Berlin, "Historical Inevitability," pp. 261–62.

42. The issue of authorship and artistic creation is no less controversial in literary and art criticism than in the social sciences. Among the best general sources for

this discussion see Rene Wellek and Austin Warren, *The Theory of Literature* (New York: Harcourt, Brace, 1956), pp. 8–42, 61–124. On the role of biography see, for example, Andre Maurois, *Aspects of Biography* (Cambridge: Cambridge University Press, 1929). In a more psychological vein see also Edward Bullough, "Mind and Medium in Art," *British Journal of Psychology* 11 (1920–21):26–46. In more sociological terms see Morris R. Cohen, "American Literary Criticism and Economic Forces," *Journal of the History of Ideas*, 1, (1940):369–74.

43. Karl Jaspers, *Man in the Modern Age* (Garden City, N.Y.: Doubleday, 1957), pp. 23–32. This is often termed a Marxist view of art and literature. See, for example, Georg Lukacs, *Beitrage zur Geschichte der Asthetik* (Berlin: Aufbau-Verlag, 1954), and his *Karl Marx und Friedrich Engels als Literaturhistoriker* (Berlin: Aufbau-Verlag, 1948). Also see G. Plekkanov, *Art and Society* (New York: International, 1936), and Leon Trotsky, *Literature and Revolution* (New York: International, 1925).

44. See the discussion on "existenal psychology" by Jean-Paul Sarte, *Being and Nothingness* (New York: Washington Square Press, 1966), pp. 712–34, and his biographical study of Baudelaire (New York: New Directions, 1950). In Marxist terms, of course, the role of the individual in artistic creation no less than in the making of society is limited. But, Marxists and non-Marxists alike often conveniently ignore Marx's own epistemological arguments, and in particular his third thesis on Feuerbach which, as V. Chernov noted, emphasizes the active role of the individual in creating the environment. For a brief review of this idea and Chernov's work see T.B. Cross, "Young Marx and Marxism: Victor Chernov's Use of the Theses on Feuerbach," *Journal of the History of Ideas*, 23 (1971):600–606.

45. Robert Caro, *The Power Broker, Robert Moses and the Fall of New York* (New York: Vintage Books, 1975).

46. Isaiah Berlin, *The Hedgehog and the Fox, An Essay on Tolstoy's View of History* (New York: Simon & Schuster, 1953), p. 29.

47. J.K. Wright, *Human Nature in Geography* (Cambridge: Harvard University Press, 1966), p. 83.

48. See, for example, the discussion by Andrew March, *The Idea of China* (New York: Praeger, 1974), pp. 23–45. We are, of course, not alone in such biased views. The "traditional" Chinese view of steppe peoples and lands was, for example, no less aimed than our own images of China. See the discussion by Marwyn S. Samuels, "Kung Tzu-chen's New Sinkiang," *Annals*, American Association of Geographers, 66 (1976):416–27.

49. Joseph Gaer, *The Legend of the Wandering Jew* (New York: Mentor, 1961).

50. Margaret Atwood, *Surfacing* (Don Mills, Ontario: PaperJacks, 1973).

51. As Adolf Hitler knew only too well, it is not the "truth-value" of impressions or images that renders them more or less objective, but their "historic-value," i.e., the extent to which they are repeated, diffused, accepted, and acquire a clientele through political, economic, legal and broadly social sanction. The "truth-value" of an image is not *in* the environment itself, but in the eye of the beholder. If the image is "exaggerated," it is only to say that someone else (a third party, or the one against whom the image may be aimed) holds a contrary view which,

through common agreement, is less "exaggerated." Whether true or false, benign or dangerous, images exist and acquire an "objective" content as they acquire a history; a history of authors in context. One essential merit of a biography of such imagery is that, by tracing the history of authorship, we expose the identity of those most responsible for the image, as well as for the landscape made in the wake of the image.

52. This is, of course, less true in cases where literary, artistic, and philosophical movements are linked directly to political and social change, as in the case of the May 4th Movement in twentieth century China. Many of the early founders and participants of revolution in China, including Ch'en Tu-hsiu, Lu Hsun, and Mao Tse-tung, were actively involved in the flourishing literary and artistic reformation of the first quarter of the century. Mao himself was, of course, no mean poet. For a general discussion see A.C. Scott, *Literature and the Arts in Twentieth Century China* (Garden City, N.Y.: Doubleday, 1963).

53. The most appropriate methodological analogue to the idea of authored landscapes in the social sciences can be identified with that which Nagel and others have described as "methodological individualism." See Nagel, *The Structure of Science*, pp. 535–46.

54. L.A. Maverick, *China A Model for Europe* (San Antonio, Texas: Paul Anderson Co., 1946), pp. 119–28.

55. Theodore Gaster, *Thespis: Ritual, Myth and Drama in the Ancient Near East* (Garden City, New York: Doubleday, 1961), pp. 17, 24–25.

56. M. Eliade, *The Sacred and the Profane* (New York: Harper & Row, 1961). Yi-fu Tuan, "Geopiety," *Geographies of the Mind*, ed. D. Lowenthal and M. Bowden (New York: Oxford University Press, 1976), pp. 11–39.

57. George Kiss, "Political Geography into Geopolitics, Recent trends in Germany," *Geographical Review* No. 4 (1942):632–45.

58. Simone Weil, *The Need for Roots* (New York: Harper & Row, 1971).

59. Hofstadter, *The Progressive Historians* pp. 47–164, 437–66.

60. This is not to suggest agreement with the Marxist notion of a direct or casual link between the doctrine of private property, capitalism, and nationalism. As Rosa Luxemburg learned to her dismay, capitalism had no monopoly on the logic or the ethic of nationalism. The national-socialism of Lenin, and especially of Stalin, was no less guilty of the worst in chauvinism than its presumed enemy. The most direct political manifestation of the doctrine of privatism is anarchism, i.e., the rejection of the state. For this reason, the authors of the socialist landscape were more opposed to the excesses of "left deviation" then to the dangers of capitalism, which is to say that they were (and remain) opposed to private property, but not the bourgeois product of nationalsim. For part of this discussion in Marxist terms see J.P. Nettl, *Rosa Luxemburg* (London: Oxford University Press, 1969), pp. 500–519.

61. It is not my purpose here to provide a definition for or theory of ideology in the making of landscape. Rather, it is to suggest that insofar as ideas or values have an impact on or expression in the landscape, some link between idea and action, some mode of transformation is required. That link is ideological in the sense

that the latter constitutes a system of beliefs, ideas or values that acquire sanction as the basis for action or behavior. It is also in that sense "religious," i.e., ideas or beliefs sanctioned by religions. For a broad discussion of religious values in the landscape see David Sopher, *The Geography of Religion*. For a sweeping discussion of the role of ideas and ideology see, among others, Talcott Parsons, *The Social System* (New York: Free Press, 1964), pp. 326–83.

62. C.B. Malone, "History of the Peking Summer Palaces Under the Ch'ing Dynasty," Illinois Studies in the Social Sciences, 19, Nos. 1–2 (1934):134–70.

63. The literature on urban design is, of course, too extensive for detailed treatment in this paper. For discussions on the contexts and the history of urban design see Paul Wheatley, *The Pivot of the Four Quarters* (Chicago: Aldine, 1971); N. Ardalan and L. Bakhtiar, *The Sense of Unity: The Sufi Tradition in Persian Architecture* (Chicago: University of Chicago Press, 1973); L. Mumford, *The City in History* (New York: Harcourt, Brace & World, 1961), and C.B. Purdonn, *The Building of Satellite Towns* (London: J.M. Dent & Sons, 1949). Though such studies contribute to an understanding of the general contexts of design, they also frequently ignore the authors of such designs. While planned cities such as Washington D.C. have been treated in terms of authorship [see, for example, J.S. Young, *The Washington Community* (New York: Columbia University Press, 1966)], to my knowledge few so-called unplanned cities have been so treated. The role of specific developers, architects, planners, and others seems only now on the verge of serious examination on the part of urban historians and geographers.

64. David Ward, "The Victorian Slum; An Enduring Myth," *Annals*, Association of American Geographers, 66 (1976):323–36.

65. George R. Stewart, *Names on the Land: A Historical Account of Place-Naming in the United States* (Boston: Houghton Mifflin, 1958).

66. It is not my purpose here to offer a theory of elitism or a definition of the role of elites in the making of landscape. For a detailed discussion of the latter see C. Wright Mills, *The Power Elite* (New York: Oxford University Press, 1956); T. Parsons, "The Distribution of Power in American Society," *World Politics* 10 (1957):123–43; Floyd Hunter, *Community Power Structure* (Garden City, N.Y.: Doubleday, 1963); A. Etzioni, *Studies in Social Change* (New York: Holt, Rinehart, & Winston, 1966), pp. 9–29; T.B. Bottomore, *Elites and Society* (Harmondsworth: Penguin Books, 1966); and T.R. Dye and I.H. Zeigler, *The Irony of Democrary, An Uncommon Introduction to American Politics* (Belmont, California: Duxbury Press, 1972).

67. E.A. Cohen, *Human Behavior in the Concentration Camp* (New York: Universal Library, 1953), pp. 115–284.

68. The datum of the obscure can become known to us so long as they remain alive, and insofar as others have bothered to record their impressions. One early effort to that end, for example, can be found in the case of M. Zborowski and E. Herzog, *Life is With the People: The Culture of the Shtetl* (New York: Schoken Books, 1951). An example closer to home is the record of migrant farmers in late 19th century America. See, for example, *Letters from the Promised Land: Swedes in America, 1840–1914*, ed. H.A. Baiton (Minneapolis: University of Minnesota Press, 1975).

Thought and Landscape

The Eye and the Mind's Eye

Yi-Fu Tuan

Landscape, from a naive viewpoint, is a sector of reality "out there." It is made up of fields and buildings. Yet it is not a bounded entity as a tree or a building is. Nor does landscape mean simply a functional or legal unit such as a farm or a township. Landscape, like culture, is elusive and difficult to describe in a phrase. What is culture and how does one delimit a culture area? The contents of culture can be itemized, although if one is meticulous the list threatens to grow to interminable length. Culture is not such a list. Landscape, likewise, is not to be defined by itemizing its parts. The parts are subsidiary clues to an integrated image. Landscape is such an image, a construct of the mind and of feeling.

Images of landscape are potentially infinite, yet they have a family likeness. This family likeness is not so much the result of shared elements in the landscape as of a common principle of organization, which evolved historically in modern Europe. Briefly stated, the principle requires landscape to be a fusion of two major perspectives—functional and moral-aesthetic. Originally, the term "landscape" referred primarily to the workaday world, to an estate or a domain. From the sixteenth century on, particularly in the Netherlands and in England, landscape acquired more and more of an aesthetic meaning; it became a genre of art.[1] Limited to the functional or utilitarian perspective, the concept of "landscape" is redundant since the more precise terms of estate and region already exist. Limited to the aesthetic perspective, "landscape" is again redundant since the word "scenery" offers greater clarity. But we do have the word "landscape" in addition to the other terms, and it is being used because we have learned to recognize a special ordering of reality for which a special word is needed.

 Landscape is an ordering of reality from different angles. It is both a vertical view and a side view. The vertical view sees landscape as domain, a work unit, or a natural system necessary to human livelihood in particular and to organic life in general; the side view sees landscape as space in which people act, or as scenery for people to contemplate. The vertical view is, as it were, objective and calculating. The farmer has to know how much land he has under each crop and how many head of cattle the pasture will support. The geographer studies the rural landscape in a similar way—that is to say, from "above"; likewise the ecologist when he looks at landscape as a natural system. The side view, in contrast, is personal, moral, and aesthetic. A person is *in* the landscape, working in the field, or he is looking out of a tenement window, from a particular spot and not from an abstract point in space.[2] If the essential character of landscape is that it combines these two views (objective and subjective), it is clear that the combination can take place only in the mind's eye. Landscape appears to us through an effort of the imagination exercised over a highly selected array of sense data. It is an achievement of the mature mind.

Children and Landscape

The abstract and practical understanding of an estate, domain or region takes time and thought; similarly, the perception of mood, of beauty and ugliness, of justice and oppression in a scene. Consider how children, as they mature, acquire increasingly complex images of landscape in the context of middle-class American culture. In the late 1950s, Frank and Elizabeth Estvan studied first- and sixth-grade pupils in the rural and urban schools of Wisconsin.[3] The children are shown four pictures in turn: village, farm, downtown, and factory. Each picture is framed in such a way and includes such elements as to represent what an adult conceives to be a typical landscape. For each picture the child is asked the open-ended question, "What does it say?" To first-graders the pictures say very little, particularly those that depict a factory and a village. Young children often fail to recognize a structured entity—a landscape—in the picture. Their first identifications are of human figures: "Here is a little girl shopping with her mother" and "There is a boy going to school." Of the physical environment young children tend to see individual elements, such as church, playground, and bicycle, rather than the larger unit that encompasses them. First-grade pupils can easily identify the farm. They are articulate in their description of it irrespective of whether they come from a country school or from a school located in the metropolis. In America knowledge about the farm, including the appeals of farm life, is inculcated at an early age. The picture showing a downtown area is also easily recognized, although the response of both city and country children to it is more ambivalent. The factory evoked the least interest and response, which is not surprising; but many first-graders fail to recognize the village and this is rather unexpected. The older pupils have no trouble identifying the pictures as representing landscape entities. In their descriptions they offer facts that are not evident in the picture, such as the regional setting of a village or farm, and how a factory functions as part of a larger system. Enthusiasm for both rural and urban scenes grows with greater understanding. The older children, for example, feel more warmly toward the farm than do the first-graders because they see more precisely what they are able to do and enjoy in such a setting.

Ecological and Sociological Appreciation

As the children mature they learn to see landscapes in an increasingly complex manner. When they look at a farmstead they are able to discern that it has a regional content, that it has special functions within a regional economy, and that the scene before them was different in the past and will change again in the future. In other words, they learn to perceive more and more with the mind's eye. Visual stimuli from the environment will increasingly trigger more trains of thought that lead lives of their own, and when the mind refocuses on the environment what it beholds is necessarily colored by its history.

　　Among the numerous ways that thought can organize and interpolate sense data, two have gained currency in recent years: the ecological and the sociological. They, can strongly affect our judgment. Consider the English countryside of hedged fields, stately houses and rustic villages. Many people like it. The landscape's charm is enhanced when viewed through ecological spectacles, but it is marred by vexing moral questions when seen from a critical sociological perspective.

The English countryside we know today owes much to the socioeconomic forces and aesthetic tastes of the eighteenth century.[4] Two hundred years of art and literature have taught us to respond to it almost as to a physical stimulus; that is to say, unreflectively. What ecological ideas add is a veneer of intellectual satisfaction to the sensory rewards. Jacquetta Hawkes invites us to see this landscape as having achieved a happy moment of balance in its long course of evolution. Before the eighteenth century, she maintains, nature still dominated the Englishman. Power was on the side of nature and it could be brutal. After the eighteenth century, as a consequence of the Industrial Revolution, the scale of power was tipped in man's favor. He proceeded to outrage the land with coal dumps and urban sprawls. For a time, however, the people were confident though not yet arrogantly so. For a time, as Hawkes put it, "Rich men and poor men knew how to raise comely buildings and to group them with instinctive grace. Town and country having grown up together to serve one another's needs enjoyed then a moment of balance."[5]

Sociological thought, unlike the ecological one, exposes the somber side of the picture. We are reminded that in Georgian England the rich had power. This power was used wisely to improve the fertility and beauty of the land, but it was also used to exploit tenant farmers and wage laborers. Whereas the abuse of land produced ugly scars that endured, the oppression of a people left few visible traces. Oppressive society and a rich beautiful countryside are fully compatible. The countryside can always look innocent, for the instruments of rural exploitation exist most conspicuously as law courts and money markets which are all in the city. Raymond Williams has shown, however, that if we look closely and with imagination, the English landscape will reveal a darker meaning. He asks us to observe the arrogant disproportion between country houses and farmers' cottages. The working farms and cottages are in the ordinary scale of human achievement. What the "great" houses do is break the scale by an act of will corresponding to their real and systematic domination of others.

> For look at the sites, the façades, the defining avenues and walls, the great iron gates and the guardian lodges. These were chosen for more than the effect

from the inside out; where so many admirers, too many of them writers, have stood and shared the view, finding its prospect delightful. They were chosen, also . . . for the other effect, from the outside looking in: a visible stamp of power, of displayed wealth and command: a social disproportion which was meant to impress and overawe.[6]

Landscape and Culture

When we look at a person we may admire his suit and personal belongings in themselves, for their functionality and handsomeness. More often we attend to the human personality for which the dress and other accouterments provide the subsidiary clues. Landscape can be evaluated in itself as beautiful or ugly, productive or infertile. On the other hand, it is also a clue to a region's human personality. Here is an example of how to see the landscape subsidiarily; that is to say, how to see *from* the landscape *to* the values and pathos of a folk. The place is the backwoods of South Carolina. The people handle snakes as part of a fervent cult. Benjamin Dunlop writes:

> Snakehandling [in the American South] is a post-industrial phenomenon, spawned by the textile mills and the rootlessness of workers flushed from the hills. . . . [The people's] religion has been jerrybuilt from the Appalachian junk-yard—the broken washing machine on the porch, the bedsprings and tin cans in the gully, the engineless Chevy on blocks in the yard. Their graves are weed-grown holes on a scraggly hillside—no hint of the dignified destitution of [the movie director] John Ford's pioneers, dust sifted to dust as the vast mesas and vaster sky loom above them. This is a backwater, not a frontier, and the faithful seem more concerned with this world than the next. Unto Caesar they render Kentucky fried chicken and suburban station wagons, unto Christ fluorescent Last Suppers and electric guitars. It is a discount-house religion, an Appalachian cargo cult, built on the rock of consumerism and hence not our contradiction but our lowest common denominator.[7]

This attempt at descriptive analysis appeals largely to the mind. There are, it is true, simple material facts in the description: one can easily take pictures of a broken washing machine on the porch, of graves on a scraggly hillside, and of the ecstatic snake-handlers. But pictures in themselves offer only superficial information. What Dunlap's vignette offers is a mental image in which visual elements of the landscape suggest, and are interwoven with, relations and values that cannot be seen. Of man's symbolic systems only the verbal kind is subtle and flexible enough to articulate such a world. We hear it said that a picture speaks a thousand words; true, but only because we have language and use words. If we were not linguistic animals, visual images could not carry even a small fraction of the meaning that they often have for us.[8]

Thoughts in a Cemetery

We can think, therefore, that we are able to see an entity called landscape. Landscape in turn evokes thought. How people act in different physical settings is fairly well known, but we know little about the thoughts run through their minds. Consider the cemetery. It is a type of landscape. It can be appraised for its utilitarian, religious, and aesthetic qualities. People visit it for a variety of reasons: a relative or a famous person is buried there and, moreover, the trees provide shade for a picnic. People may also be drawn to it for no ostensible purpose. The cemetery has a curious appeal. According to Elias Canetti, it induces a special state of mind which in turn affects how one perceives it. Canetti asks, What does someone who finds himself in a graveyard actually do? How does he move and what occupies his thoughts?

> He wanders slowly up and down between the graves, looking at this stone and that, reading the names on them and feeling drawn to some of them. Then he begins to notice what is engraved beneath the names. Here is a man who lived to be thirty-two; another, over there, died at forty-five. The visitor is older than either of them and yet they are already out of the race. But this is not the only kind of calculation which occupies the man who stands between the rows of graves. He begins to notice how long it is that some of the buried have lain there. Chronology, which is normally only used for practical purposes, suddenly acquires a vivid and meaningful life for him. All the centuries he knows of are his. Many of the things which happened during those years are known to him; he has read about them, heard people talk and experienced some of them himself. He is in a position where it would be difficult not to feel some superiority, and the natural man does feel it. But he feels more than this. As he walks among the graves he feels that he is alone. Side by side at his feet lie the unknown dead, and they are many. They cannot move, but must remain there, crowded together. He alone comes and goes as he wishes; he alone stands upright.[9]

Science and Landscape

Canetti says that a cemetery induces a special state of mind and then proceeds to describe that state of mind. From the description it is clear that Canetti does not use the word "induce" in any causal-deterministic sense; a cemetery is a special kind of place that can trigger meditation along certain lines, usually on the general theme of time and mortality.

What happens when a scientist looks at the landscape? Ideas come to his head. He develops them and in the process withdraws from the immediate sense impressions. Should he look at the landscape again, he will be able to see it in a fresh light using the developed conceptual frame as a subsidiary clue. "Time" may be such a clue. When a scientist asks how a feature in the landscape has come into being, his answer necessarily has a temporal dimension. Knowledge of this temporal dimension will affect his next encounter with the landform, provoking—perhaps—thoughts of mortality. Here is an example. A residual block of soft sandstone in New Mexico was designated a National Monument. It earned this honor because four hundred years ago some Spanish explorers had carved their names on it. Four centuries are a long time in human terms, yet in this period weather and erosion have made so little dent on the sandstone face that the signatures are preserved. A scientist will know that the sandstone block was once attached to the main scarp several miles away. In the immensity of geologic time, weathering and erosion had removed thousands of square miles of rock to produce a broad plain that now surrounds and isolates the monument. As the scientist sits in its shadow and looks across the plain to the distant scarp, he sees this panoramic space; it feels overpowering and alien to him because he perceives not only space but, given the scientific knowledge, geologic time of a dizzying remoteness.

That scientific knowledge can increase one's appreciation of landscape is not news. John Muir, for example, studied the Yosemite Valley in great detail and made original contributions to our understanding of its landforms; such knowledge only enhanced Muir's own appreciation of the valley.[10] But equally well known is the idea that scientific analysis leads to abstractions and removes us from any personal involvement with landscape or nature. For instance, the story is told of a Princeton physicist who walks about in very large boots because to him the ground is not solid and supportive, but is made up of atoms in empty space.

How does a scientist of a highly conceptual bent respond to the world? I imagine it to be something like this. Take a theoretical geographer. He begins by looking at the city, experiencing it as most people do. Very quickly, however, he abstracts data from it and builds a model out of the data. Being a good scientist, he will then test the model against abstracted data of a similar kind in other urban areas. The model is a pair of spectacles through which a scientist looks, not at the thunder showers, houses, and people, but at their quantified indices. In geographical work the indices normally lie close to the reality: for example, when I look at a rising curve of relative humidity I can almost feel the heat; likewise when I examine tables of family income I can almost envisage the kinds of houses the families live in. Because of this close relationship, a geographical model is capable of illuminating not only the indices of the real world but the real world itself, should the theoretician choose to return to it and contemplate it through the new conceptual frame. He may, of course, forgo the privilege since it is not required of him as a scientist.

The physicist has a much more difficult job integrating his theoretical knowledge with the world of daily experience. In his work the indices to reality are practically indistinguishable from that reality. A physicist can see atoms and smaller particles only as they are registered on the screens of elaborate instruments. He knows that all substance, including the ground on which he walks, consists of atoms. But his knowledge of the mathematical structure of atoms is not likely to function as a subsidiary clue illuminating the experience of walking on the hard ground: the concept and the experience are too far apart. Even if the physicist were to substitute a picture of atoms for the mathematical formula, it will remain difficult for him to amalgamate a picture of atoms with the trees and houses that he can see before him. The mental picture and the perceptual experience remain too far apart to be readily combined in the mind as a new and more vivid reality. Should, however, the physicist succeed in making the integration he is surely leading the imaginative life of an artist.

Art and Landscape

T. S. Eliot wrote: "When a poet's mind is perfectly equipped for his work, it is constantly amalgamating disparate experiences; the ordinary man's experience is chaotic, irregular, fragmentary. The latter falls in love, or reads Spinoza, and these two experiences have nothing to do with each other, or with the noise of the typewriter or the smell of cooking; in the mind of the poet these experiences are always forming new wholes."[11] Eliot has simplified to make a point. In his sense of how a poet's mind works, the ordinary man is a poet as we all are in varying degree. When we look at a landscape and see a church spire at the end of a tree-lined road, our eyes have automatically combined visual data to form a stereoscopic image, and our mind has integrated, with little conscious effort, diverse clues and experiences to give a rich meaning to that image.

Landscape, as a distinct concept sanctioned by past usage, is a fusion of disparate perspectives. We have seen earlier how it can be both a domain and a scene, both a vertical view and a side view, both functional and moral-aesthetic. To see landscape properly, different sets of data must be conjoined through an imaginative effort. Perhaps a student's attempt to see three-dimensional relief in photo interpretation provides a useful analogy. A student puts a stereoscope over a pair of aerial photographs. Looking through the instrument he is aware at first only of the flat surfaces shown on one or the other of the photographic pair. Soon, however, the data from the two sources fuse and what he then sees is three-dimensional relief—a stereo image.[12] In like manner, when a person faces the environment he may see alternatively an operational farm, a pleasant scene, and a type of social order. Should these different sets of clues amalgamate into a vividly coherent whole in his mind's eye, what he sees is landscape. But there is no instrument, no stereoscope, that will enable him to achieve the integration. He must learn by being shown fully realized examples. These examples are the works of art. A particular work may be an architect's designed environment; it may be a masterful essay in landscape description, or it may be the words of a teacher as he struggles to capture the genius of place in the classroom and in the field.

Architecture and Landscape

Architecture and literature are both achieved integrations. Both can help us to see landscapes in environments. The ways they are able to do this necessarily differ. A topographical poem affects us purely through the mind. It trains the mind to juxtapose and fuse disparate experiences. A successful building also stimulates the mind. As the lucid exemplar of how the functional and the aesthetic can be integrated, a well designed park, church, or shopping mall trains us to be aware of such syntheses in natural scenes as well as in the happy accidents of unplanned manmade environments. A poem or an essay is not itself an important element in our surrounding. By contrast, a building is. A designed landscape is indeed an all-encompassing milieu. Architecture, unlike literature, can affect our senses directly. It influences us by simply being there, bypassing the necessity to stimulate the active cooperation of the mind.

Buildings and topographical poems, insofar as they are artworks, clarify experience. They clarify different kinds of experience and encourage

We begin by looking; but to see landscape properly, different sets of data must be conjoined through imaginative effort, such as tables of family income and the kinds of houses the families live in. (P.J. Hugill)

us to select different kinds of environmental clues to attend and fuse in the imagination. Literature is made up of words that have evocative as well as analytical power; they combine subtlety with precision. Words in themselves, however, are mere vocables or marks on a page. They are abstract rather than concrete symbols and hence they can be used to reveal the dark and offensive sides of life without overpowering the listener or reader. Buildings and landscape gardens, on the other hand, are a physical environment as well as a system of symbols. While we can bear to read about suffering and death, we are less able to tolerate human agonies depicted by pictorial and plastic means. Thus, although superb architectural works illuminate experience just as novels and poems do, they focus on a narrower spectrum; they leave out the horrors and the more glaring contradictions of life.

There are, of course, exceptions. A medieval cathedral depicts the depths as well as the heights. It shows Christ bleeding on the cross and effigies of the dead lying on slabs of marble. It has dank cellars and dark corners for the confession of sins. It is able to combine these depths with the sublimity of heaven in the soaring nave and in the splendor of rose windows. A cathedral is not, however, one's ordinary environment. The setting and the rituals encourage the worshiper to distance himself from life so that he can better contemplate it unflinchingly. At the other extreme, a house designed purely for efficient living should be nearly invisible; that is to say, a resident should be as unaware of its support as he is unaware of the support of his own healthy body. The children's playground provides another case. A successfully designed playground is one in which the children are conscious only of their own kinesthetic joy and of the potential field for action. In such a setting, they are barely aware of the environmental design and equipment that make their activities possible.

Most designed environments fall somewhere between these extremes. A park serves multiple functions: its lake caters to swimming, boating, and other forms of active recreation, but the park also provides benches on which people can sit and admire the view. Shopping centers, suburbs and new towns all cater to utilitarian ends but they also aspire to something more, to images that reflect communal values and ideals, to a kind of visibility that demands attention rather than use. The communal experiences and values that an architect or city planner captures in his design is idealized. That which the architect skillfully encourages us to see is almost always something harmonious or uplifting. Thus the arts differ in how they affect our perception. Whereas literary works, paintings, and even sculpture can afford to direct our attention to the sad and offensive, buildings should not because their primary function is to support life.

Thought versus Response

The designed environment has a direct impact on human senses and feelings. The body reacts unreflectively to such basic architectural attributes as enclosure and exposure, verticality and horizontality, mass, volume, interior space and light. At a more conscious level a persons responds to the signs and symbols in the designed setting, such as the paneled walls and the deep leather chair, which speak of solid comfort and wealth. A still more conscious level exists. Imagine a connoisseur of design. When he looks at a building or a landscape garden, his relation with it is no longer simply one of "response"; he actively explores and evaluates it with his mind. He approaches it with something of the knowledge and imagination of the architect himself. The architect does not, however, normally design for the connoisseur. His landscape garden or shopping mall is not an object for thought; it is meant to be used and to affect the ordinary users at the subconscious level. This, then, is another important difference between architectural impact and the impact of literature. To illustrate the difference forcefully, consider how people might act with regard to two supreme achievements of late medieval art, Chartres Cathedral and Dante's *Divine Comedy*. Whereas a devout man ignorant of architecture can enjoy the cathedral unreflectively as a landmark or an ambience, the *Divine Comedy* will reward the reader only if he is able to make a sustained imaginative effort in line with that which enabled Dante himself to create his masterpiece.[13]

I have stressed the fact that landscape is not a given, a piece of reality that is simply there. What is given is an environment to which we respond in automatic and subconscious ways. It is important to distinguish between environment and landscape. Whereas environmental psychology can be the study of how human beings react more or less unthinkingly to the stimuli around them, landscape psychology must primarily be a study of human learning and cognition. We have seen the way children learn to integrate their different experiences and knowledge of environment into images of increasing coherence and complexity. They learn to recognize landscapes, to construe worlds, while at the same time they submit—as all organisms must—to environment's pervasive influence.

The Relevance of Landscape

If it be granted that landscape, in a general sense, is a composite feature in which elements of function and of use combine with values that transcend them, then it should be clear that landscape is not simply domain plus aesthetic appeal. Landscape is not only a village and its orderly fields, mountains and valley, but also the denuded hill country of South Carolina with bed springs and tin cans in its gullies. For here is an economically depressed region and here is also a demoralized people who seek relief in cultist fervor and in unbridled consumerism. To see this South Carolina country as landscape is to amalgamate such distinct, though related, perspectives into a single vision.

Why make the effort? What is to be gained from it? To understand the world at all we must start with the evidence of the senses. We go into the field. A physical geographer looks at the landscape and immediately proceeds to extract data of use to him in the construction of a scientific hypothesis. Landscape is for him a point of departure; he is not bound to return to it and see it with the added depth of his scientific perspective. Similarly, a concerned citizen or social scientist may go into the backwoods of Appalachia; what he sees there immediately turns his thoughts to problems of social and economic justice. The physical setting itself is of no great moment; what is important is what it tells, in visible and unmistakable signs, of human destitution and hopelessness. To such a socially concerned person it must seem frivolous to use human suffering as a subsidiary clue toward a finer appreciation of landscape.

So the question remains. Why should we want to make a landscape a focal interest? Why study it, why does it seem worthy of our close attention? Here is a tentative answer. Yearning for an ideal and humane habitat is perhaps universal. Such a habitat must be able to support a livelihood and yet cater to our moral and aesthetic nature. When we think of an ideal place in the abstract, the temptation to oversimplify and dream is well nigh irresistible. Dire consequences ensue when that dream is set prematurely in concrete. Landscape allows and even encourages us to dream. It does function as a point of departure. Yet it can anchor our attention because it has components that we can see and touch. As we first let our thoughts wander and then refocus them on the landscape, we learn to see not only how complex and various are the ways of human living but also how difficult it is to achieve anywhere a habitat consonant with the full potential of our being.

Notes

1. J. B. Jackson, "The Meaning of 'Landscape,'" *Saetryk af Kulturgeografi* No. 88, (1964):47–50.
2. E. Straus, *The Primary World of Senses* (New York: The Free Press of Glencoe, 1963), p. 318.
3. Frank J. and Elizabeth W. Estvan, *The Child's World: His Social Perception* (New York: Putnam's 1959).
4. D. Lowenthal and H. Prince, "English Landscape Tastes," *Geographical Review* 55, no. 2 (1965):188–222.
5. J. Hawkes, *A Land* (London: Cresset Press, 1951), p. 143.
6. Raymond Williams, *The Country and the City* (New York: Oxford University Press, 1973), p. 106.
7. Benjamin Dunlop, "Snakehandling in South Carolina," *New Republic*, 22 November 1975, p. 20.

8. Susan Sontag, "Photography," *New York Review of Books*, 18 October 1973, pp. 62–63.

9. Elias Canetti, *Crowds and Power* (Harmondsworth, Middlesex: Penguin Books, 1973), pp. 321–22. I have abridged Canetti's lengthy reflection on thoughts in the cemetery.

10. John Leighly, "John Muir's image of the West," *Annals*, Association of American Geographers 48, no. 4 (1958):309–18.

11. T. S. Eliot, "The Metaphysical poets," *Selected Essays* (New York: Harcourt, Brace, 1932), p. 247; quoted by Michael Polanyi and Harry Prosch, *Meaning* (Chicago: University of Chicago Press, 1975), p. 88.

12. Michael Polyani, "The Structure of Consciousness," *The Anatomy of Knowledge*, ed. Marjorie Grene, (Amherst: University of Massachusetts Press, 1969), pp. 315–16.

13. I have adapted the argument of Polanyi and Prosch, who wrote: "Works of science, engineering, and the arts are *all* achieved by the imagination. However, once a scientist has made a discovery or an engineer has produced a new mechanism, the possession of these things by others requires little effort of the imagination. *This is not the case in the arts.* The capacity of a creative artist's imaginative vision may be enormous, but it is only the vision that he imparts to his public that enables his art to live for others. Thus the meanings he can create for his public are limited by the requirement that they provide a basis for their re-creation by the imaginations of other viewers or readers." *Meaning*, p. 85.

Age and Artifact

Dilemmas of Appreciation

David Lowenthal

Awareness of the past is essential to the maintenance of purpose in life. Without it we would lack all sense of continuity, all apprehension of causality, all knowledge of our own identity. I have elsewhere explored the meanings we attribute to the past, assessed our attachment to it, and discussed some of the consequences of preserving things we regard as old, obsolete, or antiquated.[1]

Yet the past is not a fixed or immutable series of events; our interpretations of it are in constant flux. What previous groups identify and sanctify as their pasts become historical evidence about themselves. Today's past is an accumulation of mankind's memories, seen through our own generation's particular perspectives. What we know of history differs from what actually happened not merely because evidence of past events has been lost or tampered with, or because the task of sifting through it is unending, but also because the changing present continually requires new interpretations of what has taken place.

The Desire to Recover the Past

The provisional and contingent nature of history is hard to accept, for it
denies the perennial dream of an ordered and stable past. We seek refuge
from the uneasy present, the uncertain future, in recalling the good old
days, which take on a luster heightened by nostalgia. Memory highlights
selected scenes, making them so real and vivid we can scarcely believe they
do not actually survive. Virginia Woolf poignantly expresses this yearning:

> It is not possible—I often wonder—that things we have felt with great inten-
> sity have an existence independent of our minds; are in fact still in existence?
> And if so, will it not be possible, in time, that some device will be invented by
> which we can tap them? I see it—the past—as an avenue lying behind; a long
> ribbon of scenes, emotions. . . . Instead of remembering here a scene and there
> a sound, I shall fit a plug into the wall; and listen in to the past. . . . I feel that
> strong emotion must leave its trace; and it is only a question of discovering
> how we can get ourselves again attached to it, so that we shall be able to live
> our lives through from the start.[2]

> The dream of recovering the past in its entirety is itself very old. St. Augustine
> wrote of the fields and spacious palaces of my memory, where are the trea-
> sures of innumerable images, brought into it from things of all sorts perceived
> by the senses. . . . In that vast court of my memory . . . there are present
> within me, heaven, earth, sea and whatever I could think on therein . . . The
> plains, and caves, and caverns of my memory are innumerable and innumera-
> bly full of innumerable kinds of things.[3]

√Landscape was memory's most serviceable reminder, as Frances Yates has
shown. Medieval philosophers and magicians developed memory systems
that used the human landscape of the whole world.[4]

For Proust, unlike for medieval scholars, the past came unbidden
through the senses, but was never consciously to be sought. "It is a labour
in vain to attempt to recapture it. The past is hidden . . . beyond the reach
of intellect; in some material object. . . . And as for that object, it depends
on chance whether we come across it or not before we ourselves must die."

The madeleine at Combray exemplified the recovery of the past
through the senses. "When from a long-distant past nothing subsists . . . the
smell and taste of things remain poised a long time, like souls, ready to
remind us . . . and bear unfaltering, in the tiny and almost impalpable drop
of their essence, the vast structure of recollection."[5]

Such total recall is rare; most of us can no longer retrieve past scenes after we have outgrown the way we originally experienced them. But electrical stimulation in neurosurgery, one medical researcher suggests, can reawaken complete and authentic memory.

> There is a permanent record of the stream of consciousness within the brain. It is preserved in amazing detail. No man can, by voluntary effort, call this detail back to memory. But, hidden in the interpretative areas of the temporal lobes, there is a key to a mechanism that unlocks the past.[6]

Others who yearn to recover the past have sought it in the cosmos, where tangible relics might be seen or even touched. Residual physical traces, nineteenth century scholars surmised, gave potentially unlimited access to the past. They believed the record must survive; given the right techniques, nothing would elude retrieval. The mathematician Charles Babbage viewed every past event as a disturbance that reordered atomic matter. "The effects of the least material change are never cancelled, but in some way perpetuated"; hence external nature bore

> an ineffaceable, imperishable record, possibly legible even to created intelligence, of every act done, every word uttered, nay, of every wish and purpose and thought conceived by mortal man . . . so that the physical traces of our most secret sins shall last until time shall be merged in . . . eternity.[7]

For Babbage, history was preserved with a Judgmental vengeance!

The theory of relativity promises another route to the recovery of the past. Events that took place long ago on earth are only now visible in galaxies light-years away, and have yet to become manifest still farther off. From these remote purviews the whole of terrestrial history could, in theory, be seen again. "However inaccessible the past," a correspondent suggests, "every detail of life—and all other events—remain recorded in the matrix of space-time. . . . All are capable of some sort of review."[8]

The reappearance of the past in its corporeal essence is the theme of Brian Moore's *Great Victorian Collection*: the protagonist dreams of a priceless display of Victorian objects, and awakens to find that they have come into being in the parking lot of his California motel. But the collection, "a duality which exactly reproduces the originals it commemorates," rapidly deteriorates from lack of interest or faith in its reality.

> The original materials now seem false. . . . The fountain, for instance, Osler's great crystal fountain: those perfect blocks of polished glass are now mostly

dull, light in weight, dead as plastic. . . . It's the same with the silverware. The hallmarks have faded completely, so that I can't tell any more whether it's silver, or silver plate. The Staffordshire pottery has lost its glaze: it looks imitation. But in most cases it's not change but simple damage. . . . The machinery either warps or breaks down, the canvas cracks, the furniture stuffing appears, the toys don't wind, the dolls' eyes no longer move, the damask and linen have brown stains . . . the statuary has developed cracks, even in the cast-iron pieces. The musical instruments all give out false notes.[9]

The fragility of Moore's collection also characterizes actual artifacts newly brought to light. The threatened disintegration of the Punic warship excavated in the early 1970s arouses anxiety about disinterring a second one for public display; since "conservation of ancient timbers is still a universal archeological problem," pending a solution, such ships are more likely to last for posterity if left on the seabed.[10]

Routes to the Past: Memory and Artifacts

Memory and artifacts provide complementary but differing routes to the past. Freud's *Civilization and its Discontents* poses a striking analogy between them. Traces from many periods of life and of history endure both in our minds and on the ground. But what we can remember potentially includes everything that has happened, whereas what we can see of the past in physical relics is highly selective, because materials decay and because later structures on a site necessarily displace earlier ones. Hence the modern visitor to Rome, Freud suggested,

> will see the wall of Aurelian almost unchanged . . . he may perhaps trace . . . the outline of *Roma quadrata*. Of the buildings which once occupied this ancient ground-plan he will find nothing, or but meagre fragments. . . . These places are now occupied by ruins, but the ruins are not those of the early buildings themselves but of restorations.

And under the fabric of modern Rome other ancient ruins survive unseen. Suppose, however, that instead of being a tangible metropolis, Rome was a city of the mind "in which nothing once constructed had perished, and all the earlier stages of development had survived alongside the latest." In this city

> the palaces of the Caesars would still be standing on the Palatine. . . . Where the Palazzo Caffarelli stands there would also be, without this being removed,

the Temple of Jupiter Capitolinus, not merely in its latest form, moreover . . . but also in its earliest shape, when it still wore an Etruscan design. . . . Where the Coliseum stands now we could at the same time admire Nero's Golden House; on the Piazza of the Pantheon we should find not only the Pantheon of to-day as bequeathed to us by Hadrian, but on the same site also Agrippa's original edifice; indeed, the same ground would support the church of Santa Maria sopra Minerva and the old temple over which it was built. And the observer would need merely to shift the focus of his eyes, perhaps, or change his position, in order to call up a view of either the one or the other.

Unlike actual buildings, obliterated by destruction and erosion, in this imaginary city, as in the mind, "everything survives in some way or other, and is capable under certain circumstances of being brought to light again."[11] The remembered Rome is a visual amalgam of all its pasts. But Freud's metaphor also underscores the difference between memory and milieu: the present-day landscape may evoke many pasts, but can never display any period in its entirety, let alone reveal the whole of the past.

The recognition of familiar persons, places and objects enables us to reconstruct past events. A variety of aids—diaries, newspapers, books, the recollections of other people—supplement our own recall. Memory is fallible, however, and constantly altered by revision, conflation, and invention. Indeed, we cannot be sure that the entire past is not invented; perhaps the world was created just five minutes ago, as Bertrand Russell suggested, with relics and memories of a past that never was.[12] A similar view was seriously held by nineteenth century naturalists like P.H. Gosse, forced by faith in the Scriptures to believe that the earth only *appeared* to be older than six thousand years. Erosions and evolutionary successions, apparently antediluvian, had all been "put in" the act of Creation; presumably 'God hid the fossils in the rocks in order to tempt geologists into infidelity."[13] Orwell's *Nineteen Eighty-Four* carries this argument a step further:

"The earth is as old as we are, no older. How could it be older? Nothing exists except through human consciousness." "But the rocks are full of the bones of extinct animals—mammoths and mastodons and enormous reptiles which lived here long before man was ever heard of." "Have you ever seen those bones, Winston? Of course not. Nineteenth century biologists invented them. Before man there was nothing."[14]

A modern version of this proposition refers to the "discovery" of a newspaper clutched in the fossilized jaws of a 70-million-year-old tyrannosaurus, in undisturbed Cretaceous strata, proving that "the universe was in fact created at about five past nine this morning and whoever did it slipped up by leaving this copy of *The Times* lying around."[15]

Awareness of History and of Decay

We perceive the past through artifacts, physical traces, and objects in the landscape that we believe endured from earlier times, or "are old." Things persuade us of their survival in two distinct ways. One is their resemblance to, or congruence with, forms, styles, or species that are historically antique and obsolete—open field traces, vintage automobiles, classical pediments. This mode of perception might be termed antiquarian; it is essentially humanistic, requiring historical knowledge. The other is our awareness of prolonged use or decay—a worn chair, a wrinkled face, a corroded tin, an ivy-covered or mildewed wall. This mode might be termed senescent, bearing on aging: it is essentially naturalistic, requiring awareness of organic change.

Both antiquarian and senescent insights are subject to error. The Greek vase, the classical column, may be a copy or a fake; the open field pattern may be an inaccurate reconstruction; wrinkles may have been painted in, moss on the roof grown deliberately by the use of manure. Appearances of antiquity can be just as deceptive as memories of the past.

Antiquarian and senescent pastness are partly congruent, partly dissonant. Things that are historically ancient ought, we feel, to bear the marks of age. "Old buildings should look old," it has been said. "It is part of the quality we admire in them that they have their history written on their faces."[6] Even demise can seem appropriate, as with the wooden truncheons issued to the High Constables of Lichfield in the reign of William IV, but which now have death watch beetle and hence "could no longer be used."[17] On the other hand, excessive decay can spoil our appreciation of antiquity, as when buildings or statues crumble into unrecognizeability. And decay in some materials is inherently unacceptable. An assertion that "the Eiffel Tower has rusty bolts"[18] arouses a sense of disharmony: we appreciate the Eiffel Tower's historical antiquity, but this reminder of physical deterioration seems tawdry. John Earle similarly maligned seventeenth-century antiquarians for being "enamour'd of old age and wrinckles," and loving the "mouldy and worm-eaten,"[19] but in fact they admired antiquity, not decay.

Whether antiquated, decayed, or both, many physical remains do survive into the present, relics that make the remembered past tangible today. But just as awareness of history alters what we know of it, so recognizing the historicity of artifacts transforms both their significance and their appearance. To realize that something stems from the past is actively to alter it. The remainder of this essay discusses how these alterations affect the landscape. I shall focus primarily on attitudes towards things recognized as historically ancient; attitudes toward aging and decay are the subject of another essay.[20]

As the past decays, both on the ground and in our memories, we make the most of those relics that survive. When we recognize an historical object or locale, we mark it with signs, celebrate its setting, herald its existence in print, protect or restore it, recreate it in replica. The appreciation of antiquity set in train three clusters of activity: recognition and celebration, maintenance and preservation, and enrichment and enhancement. I shall explore the causes and consequences of each, and conclude with some observations on monuments, which remind us of events that took place before the monuments existed.

Effects of Recognizing and Celebrating the Past

When we identify something as old or value its antiquity we proclaim its provenance: here, we say, is something early (or original, authentic, ancient). And so we mark the site. Designation serves both to locate the antiquity on our mental map and to dissociate it from its own surroundings. It is no longer just old, but "olde." The marker emphasizes its special antiqueness by contrast with the unsignposted present-day environs, and diminishes the antique artifact's continuity with its milieu. The antiquity becomes an exhibit; we stand before it like a painting. The signpost tells us that it is in some measure contrived for our attention.[21]

If the antiquity is small and inconspicuous and the identifying marker imposing or garish, designation may obscure or even obliterate the actual relic; the marker then becomes a surrogate for the real thing. Thus the whole town of Alexandria, Minnesota, preempts the role of the Kensington Runestone found nearby. The runestone itself (a forgery, but historically evocative all the same) is tucked away in a museum; an outsized replica twelve times the size of the original dominates an outdoor park, and "the world's largest Viking" overshadows the town's main crossroads.

Certain traces of antiquity can scarcely be recognized without self-conscious contrivance. Only judicious highlighting makes one old battleground distinguishable from another; without signposts, few visitors would even be aware that they were on a historic site. So too with the birthplaces and residences of great men. Except for buildings like Monticello, whose appearance and furnishing display Jefferson's own taste and genius, structures are seldom visibly distinctive simply because they housed the famous; they are made distinctive only by means of markers—names, legends, photographs, memorabilia, guided tours. Indeed, a period montage is enough to evoke many historic figures; for tourists at Stratford-on-Avon, Tudoresque implies Shakespeare; no further specificity is needed.

Mark Twain relates an extreme instance of marking antiquity in the Bois de Boulogne. The guide showed him a tree, famed as the recipient of an assassin's bullet that missed Czar Alexander II. Neither bullet nor bullet hole could be seen, but Twain expected "the guides will point the tree out to visitors for the next 800 years, and when it decays and falls down they will put up another there and go on with the same old story just the same."[22] Perhaps, as Dean MacCannell notes, the original tree had already been replaced before Twain got there; or perhaps the tree was simply a better focus for the hawker's spiel than the hole in the ground the bullet might actually have made.[23]

The growth of tourist interest in history on the ground had encouraged the signposting of sites that "marked some event almost too meager to comprehend," one travel writer complains. " 'Near this site' (they usually begin) 'was believed to be' (they usually continue) 'the original shed where Josiah Dexter, an early settler of Dexterville, hid from four Hessians who.... ' "[24] Instead of rescuing history from obscurity, such markers drown it in trivia.

Physical markers—cairns, slabs of rock, entrance gates, directional arrows—are generally less obtrusive than written signs. Cairns and gates simply make us aware of historic features, whereas words on signposts obtrude both as physical objects and as linguistic symbols requiring appraisal and an assessment of their veracity. Thus the parish council complained that the word "castle" on a National Trust signpost at Bramber, a Norman relic in Sussex, "merited action under the Trade Description Act.... The word 'ruin' was added" as a consequence.[25]

Even the least conspicuous marker on the most dramatic site drastically alters the context and flavor of historical experience. Signs identifying the keeps, donjons, and garderobes of ancient castles, the refectories, chapels, and libraries of ruined monasteries, induce an academic frame of mind. Such markers make us consider function more than appearance,

When we identify something as old and we mark the site, we dissociate it from its surroundings, diminishing its continuity with its milieu. Sacramento, California. (P.F. Lewis)

structure rather than impression; like archeologists, we learn how things came into being and were used, rather than seeing what they look like, let alone experiencing "the *effect* of age," in John Piper's phrase.

Deploring "the recent ascendancy of the archeologist's view, and the diminishing influence of the artist," Piper inveighs against signposts:

> Those cast-iron notice boards on durable poles that punctuate the best close-range views of so many of our national shrines always seem to be whispering: "This is an object of interest. This notice board, placed where it is, proves that we, the authorities, do not regard it as an object of special beauty, otherwise we would have not defaced the view."[26]

Because signs inevitably classify what they identify, they also encourage us to *compare* historical sights, one building with another, field patterns in this landscape with that, these spears or suits of armor with those. Signposts reduce historical experience from environmental flux to the kind of order found in history books; they make the visible past feel more like the written record.

Signposts become more obtrusive each time they are seen. Below the Lake District church at Rydal, where Wordsworth's daughter attended Sunday school, a sign identifies the graveyard through which she ran to her waiting father as "Dora's Field." The historical vignette at first lends interest to the site, but palls on subsequent viewings; the sign's reminder of the Wordsworthian tableau no longer enlarges but constrains the view.

Markers occur not only in landscapes but on maps and travel literature, priming anticipation of historic sites. Many historical traces, devoid of signs on the ground, are marked only on maps. But the advance preparation, the process of checking back and forth between site and map guidebook, necessarily affect the experience. In the absence of markers on the ground we mentally erect our own: yes, there it is, I recognize it, it is in the right place, it stands out from present-day things around it.

A British Ordnance Survey decision to discontinue the regular recording of archeological sites recently underscored the effect of mapping antiquities. "Unless a site of antiquity is shown on an OS map," one critic noted, "It has no reliable authority for its existence." Cartographic insignia not only assist potential visitors, they help shield antiquities against development pressure; "without any visible traces on the surface . . . it is always difficult to persuade planning officers and developers that sites of considerable size and importance may be present below the ground."[27] Signposts on maps, no less than those on the ground, thus serve to protect as well as to identify historic sites.

The Effects of Protecting and Preserving the Past

When we cherish something old or venerable, we usually seek to preserve it from the further ravages of time, halting deterioration and extending life as long as possible.

The risk of erosion and destruction, whether natural or man-induced, may require antiquities to be fenced off, sealed away, put under glass or kept in artificially cooled and dehumidified premises. Like historical markers, these protective measures alter the conditions in which artifacts are experienced: they remove relics from the here and now, from continuity with the world around them, to an exclusive milieu. Sacrificing ambience for permanence, antiquities are uprooted from their locales and put into museums. Preservation sometimes coincides, sometimes conflicts, with esthetic, ecological, or utilitarian values. The progressive erosion of Niagara

Falls poses a typical dilemma: to keep the falls looking as painting and literature have immortalized them calls for massive and costly engineering to arrest the river's natural headward erosion. Otherwise the American Falls may disappear within a few milllenia, while the Horseshoe Falls will split into three small unspectacular drops.

Much talus already lies at the base of the falls, and hydroelectric demands episodically vary the flow of water. "These sights are not likely to make anyone feel that he is seeing or experiencing the genuine Niagara Falls. The consequent effects on tourism, a multimillion-dollar-per-year industry, could be substantial."[28]

A "Fallscape Committee" has recommended strategies based on one of three distinct aims: (1) preserving the falls as a visual monument by strengthening its structure, preventing rockfalls, and removing talus from the base; (2) making the falls an event by installing instruments to predict rockfalls and providing ringside seats for the natural spectacle, as with Old Faithful; (3) treating the falls as a continuous show by controlling the flow of water, the size of the pool below, and the amount of debris, with accompanying *son et lumière.*

The preservation of Niagara Falls thus stands in direct contradiction to natural processes. One could argue that nature should be allowed to take its course, and that people must get used to seeing the Falls change. But "Americans" image of the Falls does not change. Our ideal of a waterfall, formed by experience . . . and by images created by artists and photographers," is static and frozen.

> Paradoxically, the phenomena that the public thinks of as "natural" often require great artifice in their creation. The natural phenomenon of the Falls today has been created to a great extent by hydroelectric projects over the years. Esthetic appreciation of the Falls has been conditioned by . . . routes of tourist excursions and views from local hotel windows, as well as the efforts of artists.[29]

Other conservation criteria apply to the Old Man of the Mountain, at Franconia Notch, New Hampshire, whose granite nose would weather beyond recognition if left to itself. Surgery has preserved this 'natural" landmark since 1916, when the profile was bolted together.[30] But the Old Man, unlike Niagara Falls, is historically memorable not because of what it *is,* but because of what it *resembles.*

At a further remove from nature are Gutzon Borglum's four presidential sculptures on Mt. Rushmore, South Dakota. Few dispute the need for regular recarving to prevent erosion from disfiguring these faces in "America's Shrine of Democracy." Regular intervention is likewise necessary to preserve the Neolithic white horses outlined on Britain's chalk scarps; chalk outlines have to be redrawn and the slopes scoured so that the figures remain visible in their original locations.

Similar problems beset architectural antiquities. The Leaning Tower of Pisa must be supported; if it fell or were set upright, it would lose its unique historical value. The sculptures of the Acropolis and the Cloisters of New York's Metropolitan Museum can be protected against atmospheric pollution only be being removed from view, just as the Lascaux and the Altamira cave drawings have to be protected from corrosion by curtailing the number of visitors.

Problems are compounded when antiquarian and esthetic values conflict or when preservationists disagree among themselves. Recent disputes over Oxford's Bodleian heads and Wells Cathedral's sculptures, where the original figures had eroded so far they could scarcely be identified, exemplify the dilemma. Some sought to replace the eroded figures with new ones—as with Oxford's seventeenth century Carfax Conduit, recently restored and resettled at Nuneham Courtenay, whose new gargoyles include masks of the chairman of the Oxford Historic Buildings Trust and of the apprentice stone carver.[31] Others considered the Wells restoration "a disaster—negating the rhythm and vitiating the deeply personal nature of the original carvings."[32] In their view, the natural process of decay lent the Cathedral historical authenticity, the veneer of age reinforcing antiquarian charm.

The environs of historic sites and artifacts likewise pose difficulties. Should they be shaped to give relics an appropriate ambience, or to make them more viewable? Should surroundings be restored as they originally were? Or should they serve other purposes, such as interpretation, recreation, or car parking? The debate over Rodia's Towers, in the Los Angeles ghetto of Watts, illustrates these issues. Simon Rodia, a semiliterate Italian immigrant obsessed by a wish to be remembered, created his towers between 1921 and 1954 out of bottles, cans, cement, sea shells, broken dishes, and bits of old iron. The towers were saved, first from public ignorance and neglect, then from city officials who feared they might collapse: were the Leaning Tower of Pisa in Los Angeles, it would be torn down for the sake of public safety. But Rodia's towers were finally pronounced safe when they withstood a test drag of a 70,000-pound load.

The towers subsequently embroiled preservationists of various persuasions. Some sought to make them a major tourist attraction, others to revivify a Rodia-centered neighborhood. "You could write a little fairy tale about it," a critic of this plan warned:

> Here are some lovely strange towers, and some bad people want to tear them down and other bad people are already wrecking them. And these nice people come along and save them from all the bad people. Then they fix them up. And pretty soon they put in a house for art classes, and then a teen-age theatre. . . . And then, one day, they're about to put in a swimming pool or an addition to the community center, and one of the planners says to another one, "What are those dilapidated old towers over there? Why not get rid of them to get some room?" So the nice people get their bulldozers and they tear the towers down."[33]

Still other preservationists wanted to turn the clock back to 1921, when Rodia began his work: neighborhood residents "would be allowed to remain in their homes as long as they agreed not to park new cars in the driveways or put television aerials on their roofs."[34]

The difficulty with preserving environs, natural or cultural, is that like Niagara Falls they are always in flux. "Because forest growth and other vegetative ground cover is not static, we cannot exactly duplicate historic conditions," counsels a park preserver. We can only approximate "the general visual aspect of the historic period." National Parks' treatment of battle-fields illustrates the need for compromise:

> At Gettysburg . . . we can insure that there will always be a copse of trees at the Bloody Angle, although the trees may not always be the exact size of those that were there in 1863. . . . We can see that the forest does not advance into what at the historic moment were open meadows and thus interfere with vistas crucial to an understanding of the Gettysburg battle. But . . . we cannot duplicate the trenches at Petersburg or the earthen fortifications at Yorktown. Here, for obvious practical reasons, we must compromise with nature and cover with sod what strict historical accuracy would dictate to be exposed raw earth.[35]

Open-air environs of historic sites often look wrong because their custodians are reluctant to alter Mother Nature. "Officials afflicted by . . . the naturalist syndrome tend to equate historical resources with old buildings," as Robert Utley says. They have great difficulty seeing the trees and bushes and grass and rocks as historic too. Thus they rarely can bring themselves to tamper with what God hath wrought, even though in historical parks the goal is to display what man hath wrought—and the setting in which he wrought it."

The preservation of nature requires an appropriate time-scale or a focus on process rather than form. Many natural scenes are episodic. One can envisage the problems inherent in keeping Old Faithful on time, ensuring enough wave for the tidal bore of the Severn, guaranteeing the annual reappearance of autumn colors, or replicating the familiar and precious experience of sunrise and sunset.

Historic landscapes are harder to protect than buildings partly because few view "natural" features as historical. The Department of the Environment in Britain has rejected efforts to broaden the definition of "ancient monuments" to include a Roman road, a Saxon boundary hedge, and the like as being "inappropriate";[37] it would, indeed, be a daunting task to look after so numerous and multifarious a collection of sites.

Archeological sites pose preservation problems for both artifacts and environs. Excavation may derange antiquities that viewers value because they have been so long essentially undisturbed. The knowledge of prehistory derived from digging should be weighed against the sense of the past that comes from seeing untouched ancient landscapes.[38] At the same time, earth-moving technology makes it increasingly difficult to protect sites against development and farming; archeological features that would once have taken several seasons to destroy can now be demolished in half a day. Yet strict conservation carries its own risks: protected archeological sites attract rabbits displaced from newly plowed downland and grassland, their traditional habitat (rabbits burrowing can, obviously, cause widespread damage in ancient earthworks).[39]

The Effects of Enhancing the Past

When the recognizable past falls short of our historical ideals, we remold it to our desires. Old landscapes, buildings, and artifacts are decorated, purified, homogenized, emulated, copied. Some additions aim to make the past more visible, more apparent; like markers, they advertise the antiquarian landscape or artifact. Other additions seek to make an eroded and fragmentary past more coherent or evocative, like Sir Arthur Evans's reconstruction of the palace of Knossos in Crete. Evans felt that "reconstitution" was essential to give the excavated palace a comprehensible aura of antiquity.[40]

He stopped short of recreating a lived-in city; making Knossos a present-day place would have vitiated the impression of age.

Total restoration subverts historical awareness. The newly-minted completeness of Williamsburg and Old Sturbridge Village makes it difficult to imagine oneself back in the past. Old-time activities like candle-dipping, sitting in the stocks, and horse-shoeing lend an historic aura that fails to endure; as a 15-year-old said when she doffed her seventeenth century costume at Clarke Hall, a 300-year-old farm where visitors engage in antiquated tasks, "I can feel it all slipping away."[41] The breeding of medieval striped piglets at Butser in Hampshire, the preservation of endangered relic species by the Rare Breeds Survival Trust, the building and testing of Iron Age tools and buildings at Lejre in Denmark, do add to our understanding of how life was lived long ago. But living for a year under Iron Age conditions, as ten young men and women recently did in Dorset, did more to dehistoricize the remote past than to make it vivid, let alone authentic.[42]

Where the past becomes highly popular, ancient sites lose historical specificity in a romantic blur catering for any cult of the ancient. During midsummer week, over "fifty different groups come to Stonehenge," one participant explains, "because in an unstable world it is proper that the people should look for stability in the past."[43] The "Druids" who disport there on midsummer's eve owe their origins to eighteenth century antiquarianism rather than to any continuity with their priestly prototypes—who themselves never used Stonehenge, already a ruin by the time they reached Britain.[44] Yet some felt that the "Druids'" rites—they "had lit the outer ring . . . to make the monument appear to float on the plain like an unearthly silent starship"—were "wholly in keeping with the timeless magic of the place."[45]

Giving tangible form to what we feel ought to have been, as opposed to what the surviving remains suggest, prompts other alterations of the past. Some restorers genuinely believe that they are rectifying the historical record, correcting the defects of differential erosion, replacing evidence obliterated by time or by enemy action. But such efforts, however scrupulous, like Victorian Gothic inevitably convey the flavor of their own day. Other historical remains are selectively preserved—some embellished, some eliminated—out of deliberate partiality: the desire to celebrate noble and virtuous episodes and to forget the unseemly, as in the neglect of the traitor Benedict Arnold by conservers of his Connecticut birthplace; of robber baron Jay Gould in Tarrytown, New York; and of the Gold Rush prostitutes of Kern County, California, to whom nostalgic admirers erected a commemorative plaque that outraged city fathers forthwith dismantled. Poles are enhancing their ethnic past by selective attention, restoring Slavonic churches but leaving Teutonic churches to decay.

A desire for profit or for pedagogy makes remnants from the past more clustered, uniform, and homogeneous than if untouched. Moving scattered survivals into one historic precinct, confining attention to relics from a single epoch, enhances their impact on the viewer. In sites that attract large numbers of visitors, both safety and visibility require concentrating historic features. But spatial and temporal purity render historic areas static and lifeless. Milieus that deliberately exclude the remote and the more immediate past—a tastefully restored Colonial village, a ghost mining town, a Gay Nineties downtown street—are as sterile and as atypical of their own periods as a brand new subdivision today.[46]

Rearranging the past reinforces the tendency of markers and boundaries to segregate past from present, to divorce yesteryear's rare survivals from today's ordinary milieus. Museums are extreme instances of such segregation. They contain relics removed from their places of origin and gathered together in wholly contrived surroundings. Even the most sophisticated period montage conveys a simulated flavor, if only because "the backgrounds are inevitably anachronistic—classical antiquities at Bath, medieval treasures at Westminster Abbey, Renaissance paintings at the National Gallery."[47] But the museumized past, however contrived it may look, has enormously enhanced our historical knowledge and awareness, while preserving antique treasures from erosion, pillage, and neglect.

What is historically appropriate or profitable in one era will not suit another: different epochs, aspects of life, and modes of presentation succeed one another in popularity and hence in public concern for enhancement. A generation ago, "Gone with the Wind" portrayed a South considered virtually synonymous with plantation Great Houses; today replica slave quarters, significant for black historical consciousness and suiting the folk history ethos, teach quite different lessons in the same Southland. Each shift in historical perspective requires highlighting some parts of the past and playing down others.

Partiality toward origins and beginnings also affects which artifacts are preserved and how they are displayed. This concern takes two forms: a competitive urge to push back the beginnings of history (national, local, ethnic, or whatever) to a time as remote as possible; and a preference for ancient as opposed to recent forms and relics. The first theme is exemplified in Piltdown Man, a contrivance that long deceived British paleontologists, who were eager to show that mankind in Britain antedated Neanderthals from the Continent;[48] in impassioned attachments to bogus pre-Columbian Viking relics in North America;[49] and in present-day black African claims to a glorious civilization previous to any north of the Sahara.[50]

The second theme, the preference for ancient artifacts, involves a familiar archeological dilemma. When multiple pasts coexist on a site, which strata should be preserved and which sacrificed? The decision often reflects a bias toward the primordial. To explore the origins or earliest known roots of a culture or technological form, the archeologist must displace later deposits and artifacts. But partiality toward the primordial transcends this need. Relics and traces of the earliest forms and occupances are commonly expected to yield insights both different from, and greater than, subsequent ones.

This view embodies two erroneous assumptions. One is that the oldest extant relics necessarily have the greatest bearings on beginnings, whereas things in fact begin at all times throughout history. The other is that a knowledge of origins reveals more than other kinds of history, whereas it is in fact no less important to understand ongoing processes.

The claims of antiquity threatened to destroy the continuity of visible history in two recent British instances. At Avebury, some archeologists wished to highlight the megalithic circles by removing later "intrusions," including a medieval tithe barn athwart one of the embankments. At Bury St. Edmunds, historical purists wished to remove eighteenth century houses built into the medieval town walls. Both efforts failed; the collage of successive uses over time overrode the exclusive claims of the earlier remains.

Older forms are also thought superior when subsequent artifacts are felt to have declined from primitive virtue. A preference for primacy animated restorers who stripped churches of Baroque ornaments, leaving only the bare Romanesque or Gothic naves and aisles.

Restorers who regard accretions as excrescences are apt to suppose that earlier structures were more austere than later ones.[51] But previous epochs were no less given to decoration than subsequent ones; ancient remains only seem simpler because embellishments were more perishable than the structures that housed them.

A desire to figure in history impels others to alter historic scenes. Centuries of graffiti, left by visitors bent on nominal immortality, deface ancient churches, castles, and ruins. Renaissance scribblers on the walls of the Catacombs felt they were extending Classical continuity.[52] The temptation seems irresistible; even those who scold follow suit, like the historical painter Robert Ker Porter, who carved his name on the great portal of Persepolis along side the names of Pietro della Valle, Niebuhr, Cardin and Rich.[53]

Canadian troops in the Second World War disfigured the James Paine bridge at Brocket Park, Hertfordshire, "by cutting their names, with addresses in Canada, and personal numbers, all complete and inches deep—the vandals. "Yet," adds Lees-Milne, "I thought what an interesting memorial this will be thought in years to come and quite traditional, like . . . Byron's name on the temple at Sunium."[54]

If not the name of the carver, then the timeworn quality of the embossed surface embellishes the historical aura for later observers. Harrow School's heavily inscribed vintage desks recently sold in New York for $250 each.[55] The authenticity of Shakespeare's chair at Stratford may be questionable, but its erosion by famous eighteenth and nineteenth century visitors, who sliced off pieces of it "according to the Custom," is beyond doubt.[56] It was the footprints of Hollywood's great in the concrete outside Grauman's Chinese Theatre that gave that place its historic fame.

A like historical spirit animated Robert Rauschenberg's acquisition of a de Kooning drawing for the sole purpose of destroying it. Like everything in the landscape, works of art undergo continuous erosion; deliberate eradication, Rauschenberg felt, would show historical appreciation. Persuading de Kooning his argument was valid, Rauschenberg was given one of the artist's finer drawings, and methodically erased it. "It wasn't easy, by any means," Rauschenberg recalled. "The drawing was done with a hard line, and its was greasy, too, so I had to work very hard on it, using every sort of eraser. But in the end . . . I *liked* the result. I felt it was a legitimate work of art, created by the technique of erasing."[57] Underneath the drawing, its original lines just detectable on the white paper, the inscription reads:

ERASED DE KOONING DRAWING
ROBERT RAUSCHENBERG
1953

The keynote exhibit in a 1976 show at the Modern Museum of Art in New York, "Erased de Kooning" was praised as "the first work with an exclusively art-historical content and produced exclusively for art historians."[58]

Between these three types of activities—marking, protecting, enhancing—no firm line can be drawn. As tangible relics become desirable, identification inevitably leads to a need for protection, and protection to embellishment. Indeed, it is hardly possible to mark a site without being concerned about safekeeping, even against inadvertent intrusion by those ignorant of its historical importance. Likewise we can scarcely avoid enhancing things we wish to preserve. The very process of preservation—slowing down deterioration or guarding against accident—changes the look and feel, if not the form and substance, of protected sites or artifacts. Any widely appreciated trace of history increasingly risks transformation.

How Monuments Affect our Awareness of the Past

Monuments and memorials locate the remembered or imagined past in the present landscape. Their function is not to preserve the past but to recall and celebrate it. They seldom point the way to historic localities or structures, but stand instead as evocative reminders of some epoch's splendor, some persons's power or genius, some unique historical event.

Memorials thus occupy a place in the landscape wholly unlike that of the signposts and other embellishments described above. Enhancements alter or add to history on the site; memorials adorn localities that may have no connection with the celebrated person or event. Memorials are no more tied to date than to place; few monuments are of the same vintage as the event or person they commemorate. In time, old memorials may become historical landmarks in their own right; but they are seldom initially intended as such. On the contrary, some structures only become monuments *after* a prolonged existence. Thus eighteenth century Radford Lodge in Plymouth has merit largely because it leads to what was Radford House, the sixteenth century structure (demolished 1937) in which Sir Francis Drake stored much of his Spanish treasure and whence Sir Walter Raleigh was taken to the Tower.[59]

Memorials designed to impress the viewer often dominate their environs. The Albert Memorial is a familiar part of the Londoner's everyday scene; the Anzac cenotaph in Melbourne is a site of steady pilgrimage; a major element in the landscape, each serves no purpose other than to remind us of that man, this war.

Innumerable statues of departed rulers and leaders, poets and preachers, grace places that may be half a world away from where their prototypes set foot. Julius Caesar was commemorated throughout the Roman Empire; George Washington stands, or more frequently bestrides a horse, in countless American city squares and village greens; Queen Victoria looms over busy

Cemeteries matter less as repositories for the dead than as fields of remembrance for the living. With the passage of time memorials acquire an antiquarian look, their form and decor evoke past epochs, and the landscape of commemoration becomes conjoined with the landscape of the past. (above) Oakwood Cemetery, Syracuse, New York (Oakwood-Morningside Cemetery Association); (below) El Paso, Texas. (Texas Highway Department)

intersections in places as remote from Buckingham Palace as Benares and Berbice. These statues compel us to reflect on the past, often so insistently that political rebels pull them down as symbols of tyranny. Such was the fate of the Vendôme, which cost Parisians almost as much to destroy as to erect.

Most memorials are in cemeteries. Many initially served to mark the graves beneath them, but the marking function is no longer consequential once bodies have moldered into dust or have been removed to make way for others. In any case, cemeteries mark no significant event in most people's lives; we seldom die in them, but are simply put there for memorial convenience.[60] Cemeteries matter less as repositories for the dead than as fields of remembrance for the living; the unmarked grave goes unseen.

As landscape features, cemeteries are assemblages of personal memorials. Often the collective quality of memorialization stands out: in military cemeteries, the massed and uniform crosses, the anonymity of the graves, evoke not individual soldiers but the Great War in which they died. Disused and abandoned cemeteries all seem more and more collective, because those interred there matter less and less to the living. Tombstones and headstones refer increasingly to a common past rather than to the specific persons whose names appear on them. Meanwhile the cemetery takes on secondary historical characteristics. The memorials acquire an antiquarian look and a patina of age; their form and decor, the nature and calligraphy of their engraved messages, evoke past epochs. This antiquarian effect is seldom intended, either by those who designed the cemetery or by thsoe who buried their dead there. But it ultimately conjoins the landscape of commemoration with the landscape of the past. No longer just a set of monuments to the departed, the cemetery becomes a relic in its own right.

Monuments and memorials thus add to our awareness of the past. They share with relics that are marked, preserved, and enhanced a distinctiveness that comes from referring to time past. Memorials and monuments differ from other forms of historic appreciation, I have suggested, both in being subsequent to the times they point to and in their freedom from ties to locale. But by no means are all memorials so dissociated. Crowds of pilgrims venerate the burials sites of, for example, John F. Kennedy, Napoleon, and Elizabeth Barrett Browning, whose tangible remains—unlike those of Lenin or Jeremy Bentham—matter less than their final resting place. On the contrary, actual relic features continue to be revered when they are moved from their original sites—the Parthenon frieze to the British Museum, poets to Westminster Abbey, London Bridge to Arizona—no matter how much the removal violates historical integrity.

Awareness of the tangible past thus comes from a melange of traces, ruins, and artifacts. Some are fixed forever to particular locales—Hadrian's Wall, the Great Wall of China, Gettysburg. Others can only be seen when taken out of their original locations—Indian arrowheads, Egyptian mummies—or, like the Anzac cenotaph, commemorate events that took place far away. Between these extremes countless relics exhibit varying attachments to locality and landscape. In being both mobile and attached, these survivals from the past resemble ourselves: we too move about the world to protect and enhance our interests, yet are also rooted in particular locales whose history commingles with our own.

Conclusion: To Value the Past is to Alter the Past

The past, like the present, is always in flux. When we identify, preserve, enhance, or commemorate surviving artifacts and landscapes, we affect the very nature of the past, altering its meaning and significance for every generation in every place. Paradoxically, growing interest in the past threatens its visible remains and tends to vitiate their significance, especially where attention is concentrated on a few sacred places. In such historic precincts, writes Kevin Lynch, "everyday activities progressively decamp, leaving behind a graveyard of artifacts; tourist volume swells, making it impossible to maintain the site 'the way it was'; what is saved is so self-contained in time as to be only peculiar or quaint."[61]

Inveighing against tampering with ancient buildings, Ruskin a century ago contrasted self-conscious preservation in England with the harmonious conjunction of past and present on the Continent:

> Abroad, a building of the eighth or tenth century stands ruinous in the open street; the children play around it, and peasants heap their corn in it, the buildings of yesterday nestle about it, and fit their new stones into its rents, and tremble in sympathy as it trembles. No one wonders at it, or thinks of it as separate, and of another time; we feel the ancient world to be a real thing, and one with the new.... We, in England, have our new street, our new inn, our green shaven lawn, and our piece of ruin emergent from it—a mere *specimen* of the middle ages put on a bit of velvet carpet to be shown, which, but for its size, might as well be on the museum shelf at once, under cover. But, on the Continent, the links are unbroken between past and present.[62]

But the links are unbroken only so long as no one realizes how unlike the present the tangible past is. Imagine the effects of a visit to that eighth-century ruined French castle of even a small fraction of Ruskin's followers. *They* would wonder at it, think it of another time, sketch and photograph it, and transform the children into pimps and picturesque likenesses on Kodachrome. Villagers would provide lodgings and souvenirs—post cards and replicas of the castle—and mark the way with signposts. Publicity would swell the press of visitors and require the government to fence off the castle, station guards, and charge admission to defray these costs. Conscious appreciation of the tangible past always sets in motion forces that alter it.

Such transformations are potentially dangerous, for they can falsify and destroy the real past. Yet they are also potentially beneficial, for they help free us from conscious or unconscious dependence on a mythical past. It may or may not be wrong to alter the past; but it is inevitable. As Orwell feared, "who controls the present controls the past."[63] I have tried to show that this is true, not only for totalitarian societies that deliberately expunge and alter the past, but for all human beings.

We do require reminders of our heritage in our memory, our literature, and our landscapes. But advocates of preservation who abjure us to save unaltered as much as we can fight a losing battle, for even to appreciate the past is to transform it. Every trace of the past is a testament not only to its initiators but to its inheritors, not only to the spirit of the past, but to the perspectives of the present.

Notes

1. David Lowenthal, "Past Time, Present Place: Landscape and Memory," *Geographical Review* 65, (1976):1–36.
2. Virginia Woolf, *Moments of Being* (London: Sussex University Press, 1976), p. 67.
3. *The Confessions of St. Augustine*, VIII, 12, 14 XVII, 26 (New York: Grosset & Dunlap, n.d.), pp. 214–15, 223–24.
4. Frances A. Yates, *The Art of Memory* (Harmondsworth, Middlesex: Penguin Books, 1969).
5. Marcel Proust, *Swann's Way* Vol. 1, (New York: Henry Holt, 1925), pp. 57–58, 61.
6. Wilder Penfield, "Some Mechanisms of Consciousness Discovered during Electrical Stimulation of the Brain," quoted in Justin O'Brien, "Proust Confirmed by Neuro-Surgery," *Publications of the Modern Language Association* 85 (1970):295–97.
7. Charles Babbage, *The Ninth Bridgewater Treatise: A Fragment* (London: 1838), paraphrased in George Perkins Marsh, *Man and Nature; or, Physical Geography as Modified by Human Action* (New York: Charles Scribner, 1862), pp. 548–49n.
8. Michael Kirsch, cited in Peter Laurie, "About Mortality in Amber," *New Scientist* 3 April 1975, p. 37. This version is anticipated by George Perkins Marsh in "The

Study of Nature," *Christian Examiner* 68 (1968):33–62 (where the paraphrase of Babbage first appears): "Beings with faculties analogous to ours might choose at pleasure stations from which they could follow, with bodily organs, the lost history of our globe, and even now witness the rearing of the pyramids, the founding of the walls of Rome, the battles of an Alexander, the triumphs of a Caesar, or the inauguration of a Washington!" (p. 41).

9. Brian Moore, *The Great Victorian Collection* (London: Jonathan Cape, 1975), pp. 203–13.

10. R.E. Sutcliffe et al., for the Council for Nautical Archaeology, "Saving the Cheops Ship," *Times* (London), 9 June 1978, p. 15; see also Edward Gueritz et al., "Saving a Punic Warship," *Times* (London), 24 June 1978, p. 15.

11. Sigmund Freud, *Civilization and Its Discontents* (London: Hogarth Press, 1946), pp. 15–18.

12. Bertrand Russell, *The Analysis of Mind* (London: George Allen & Unwin, 1921), p. 159.

13. Quoted in Elizabeth Hardwick, "Piety and Politics," *New York Review*, 5 Aug. 1976, pp. 22–28, ref. p. 22.

14. George Orwell, *Nineteen Eighty-Four* (Harmondsworth, Middlesex: Penguin Books, 1954), p. 213.

15. Karl Sabbagh, "Weekend Competition," *New Statesman*, 11 August 1967, p. 183.

16. "St. Paul's: Black or White?" *Architectural Review* 135 (1964):243–45; ref. p. 244.

17. *Daily Express* quoted in "This England," *New Statesman* 95 (1978):813.

18. Dean MacCannell, *The Tourist: A New Theory of the Leisure Class* (New York: Schocken Books, 1976), p. 133.

19. John Earle, "An Antiquary," in *Micro-cosmosgraphie or, A Piece of the World Discovered in Essayes and Characters* (London: Golden Cockerel Press, 1928; orig. pub. 1628), p. 13.

20. David Lowenthal, "Nature, Age, and Beauty," in J.H. Appleton, ed. (untitled; in press).

21. John Summerson makes an analogous point with regard to ruins: "Once a ruin is preserved, the ruination seems no longer a work of pure accident. The 'owner' comes back into the picture, especially if he fences the site and still more if he charges an admission fee." *Heavenly Mansions, and Other Essays on Architecture* (New York: W.W. Norton, 1963), p. 237.

22. Mark Twain, *The Innocents Abroad or the New Pilgrim's Progress* (New York: New American Library, 1966; orig. pub. 1869), p. 101.

23. MacCannell, *Tourist*, p. 128.

24. William Zinsser, "Letter from Home," *New York Times*, 18 August 1977, Sec. C, p. 16.

25. *Daily Telegraph*, quoted in "This England," *New Statesman*, 1 March 1974, p. 289.

26. John Piper, *Buildings and Prospects* (London: Architectural Press, 1948), pp. 96, 102.

27. Graham Webster, "Mapping Buried History," *Times* (London), 31 October 1977, p. 13.

28. Martin H. Krieger, "What's Wrong with Plastic Trees?" *Science* 179 (1973):446–55; ref. p. 447.

29. Ibid., p. 448. See also Rayner Banham, "Goat Island Story," *New Society*, 13 January 1977, pp. 72–73.

30. John Kifner, "Franconia Notch: Highway Battle Site," *New York Times*, 29 July 1974, p. 25.

31. *Observer Magazine*, 26 September 1976, p. 7

32. John Schofield et al., "Wells Cathedral Sculptures," *The Times*, 8 February 1973, p. 17.

33. Calvin Trillin, "I Know I Want to Do Something," *New Yorker*, 29 May 1965, pp. 72–120, ref. pp. 108–9.

34. Ibid., p. 109.

35. Robert M. Utley, "A Preservation Ideal," *Historic Preservation* 28, no. 2 (April–June 1976):40–44, ref. pp. 42–43.

36. Ibid., p. 43.

37. *Commons, Open Spaces and Footpaths Preservation Society Journal* 20, no. 2 (1978):5.

38. Peter J. Fowler, *Approaches to Archeology* (London: Adam and Charles Black, 1977), chapters 1 and 6.

39. Roger Milne, "The Living Threat to Britain's Dead Past," *New Scientist*, 6 October 1977, p. 6.

40. Rose Macaulay, *Pleasure of Ruins* (New York: Walker, 1953), pp. 111–12, and Gerald Cadogan, *Palaces of Minoan Crete* (London: Barrie and Jenkins, 1975), p. 51, offers assessments of Evan's reconstructions.

41. Quoted in Ivor Smullen, "Visiting the Past," *Observer Magazine* (London), 16 October 1977, p. 75.

42. John Coles, *Archeology by Experiment* (London: Hutchinson, 1973), R.W. Apple, Jr., "Ten Britons Find Iron Age Life Tests Mettle," *International Herald Tribune*, 6 March 1978, p. 4.

43. Sid Rawle, "Solstice Invasion of Stonehenge," *Times* (London), 28 June 1978, p. 17.

44. Graham Webster, "Protecting Stonehenge," *Times* (London), 22 June 1978, p. 17.

45. James Mitchell, "Druids at Stonehenge," *Times* (London), 30 June 1978, p. 19.

46. John B. Jackson, " 'Sterile Restorations' Cannot Replace a Sense of the Stream of Time," *Landscape Architecture* 67 (1977):194.

47. Jane Carpenter, "Case of the Hidden Exhibits," *Times*, 25 June 1978, p. 15.

48. Ronald Millar, *The Piltdown Men: A Case of Archeological Fraud* (Frogmore, St. Albans: Paladin, 1974), pp. 97–98, 112–14, 193.

49. E.g., Barry Fell, *America BC*.

50. Charles Harrison, "Africans Unite to Study their Heritage," *Times* (London), 4 April 1978, p. 9, details the first general assembly of the Organization for Museums, Monuments and Sites of Africa.

51. Nikolaus Pevsner, "Scrape and Anti-scrape" and Jane Fawcett, "A Restoration Tragedy: Cathedrals in the Eighteenth and Nineteenth Centuries," *The Future of the Past: Attitudes to Conservation 1147–1974*, ed. Jane Fawcett (London: Thames and Hudson, 1976), pp. 35–53 and 75–115, respectively.

52. Erwin Panofsky, *Renaissance and Renascence in Western Art* (London: Paladin, 1970), p. 173.

53. Macaulay, *Pleasure of Ruins*, p. 146.

54. James Lees-Milne, *Ancestral Voices* (London: Chatto & Windus, 1975), p. 5.

55. Roderick Gilchrist, "The Latest Craze—a Harrow Desk," *Daily Mail* (London), 25 January 1977, p. 2; Shawn G. Kennedy, "Bit of Eton Tradition for Sale in New York," *New York Times*, 10 January 1976.

56. Richard H. Howland, "Travelers to Olympus," *With Heritage So Rich* (New York: Random House, 1966), pp. 147–50, ref. p. 150.

57. Calvin Tomkins, "Moving Out," *New Yorker*, 29 February 1964, pp. 39–85, ref. p. 59.

58. Harold Rosenberg, "American Drawing and the Academy of the Erased de Kooning," *New Yorker*, 22 March 1976, pp. 106–10, ref. p. 108.

59. "Vanishing Britain," *Times* (London) 31 May 1978, p. 16.

60. See Wilbur Zelinsky, "Unearthly Delights: Cemetery Names and the Map of the Changing American Afterworld," *Geographies of the Mind*, ed. David Lowenthal and Martyn J. Bowden (New York: Oxford University Press, 1975), pp. 171–95.

61. Kevin Lynch, *What Time Is This Place?* (Cambridge, Mass.: MIT Press, 1972), p. 237.

62. John Ruskin, *Modern Painters*, part V, chap. 1, nos. 3–5 (New York: John Wiley & Sons, 1886), IV, 3–4.

63. Orwell, *Nineteen Eighty-Four*, p. 31.

The Landscape of Home

Myth, Experience, Social Meaning

David E. Sopher

In the archetypal tale of the wanderer's return, the long, harrowing, often despairing voyage home ends at last in a paradox: Odysseus, left asleep by friendly sailors on the strand of his beloved Ithaca, wakes—and does not recognize the landscape of home.[1] The poet's explanation of this happening is the merest artifice. Instead, we may choose to construe the event as standing for the discrepancy between memory's image and current actuality. I want to explore this discrepancy between image and reality, puzzling out the ambiguities that beset the image of home.

What I shall be laying out here will be scarcely more than an agenda: an outline of questions about the meaning of home and the nature of its landscapes. These questions are raised in the context of a small literature, of recent date, in the field of geography, the core of which is made up of several finely wrought essays by Yi-fu Tuan.[2] Tuan writes from what he calls a phenomenological or experiential perspective. Mine is the comparativist one that is always implied in social science, a view that is on the whole less likely to be colored, however inadvertently, by idiosyncrasy and parochialism.

A Note on Semantics

Let us begin with the concept of "home" itself. The rich meaning of the
English lexical symbol is virtually untranslatable into most other languages. The distinguishing characteristic of the English word, which it
shares to a certain extent with its equivalent in other Germanic languages,
is the enormous extension of scale that it incorporates and has done
throughout its history. It can refer with equal ease to house, land, village,
city, district, country, or, indeed, the world. It transmits the sentimental
associations of one scale to all the others in a way that the Romance Languages, for example, can not. To speak of "hometown" or "homeland", in
which the scale is made specific as a matter of convenience, is to transfer
the same warm feelings of security and familiarity that we experience at the
scale of the family dwelling to a city or a country, in a way that speakers of
French and Spanish do not. In those languages, and many others, house and
home are one and the same: to be at home is literally to be "inside one's
house." The Romance word for "house" then takes on some of the warmth
associated with "home" in English, but it remains a symbol for a firmly
bounded and enclosed space, which "home" is not.

In *La poétique de l'espace* ("The Poetics of Space"), an essay in the
phenomenological manner, Gaston Bachelard writes about the poetic images of familiar friendly spaces, giving special attention to the house and
the experience of it in childhood.[3] Reading it in the English translation, in
which *la maison* of the original is rendered by the English word "house," we
are left vaguely disquieted. So much of the intimate, lived-in character of
domestic space in Bachelard's memories and in the poems he quotes seems
to have been drained away in the English; we feel that the "house" is
empty, as if the movers have already left with the furniture. On the other
hand, it will not do to insert the semantics of home where Bachelard has
the notion of a duality of inside/outside, as Douglas Porteous has done
recently.[4] His assertion that "the fundamental dichotomy in geographical
space is between home and non-home" sounds exceedingly odd because the
English word does not imply the existence of a sharp boundary or a fixed
scale.

Beyond the inadequacies of translation, subtle cultural differences
appear. Among the many quotations in Bachelard, selected from a wide
range of French poets and writers, the one that articulates explicitly the
common English sentiment of "love of home" is indeed a passage from a
book in English.[5] English is indefinite not only as to the extent of home, but

also as to its content, so that it is in fact understood to incorporate family— or at other scales, kin, neighbor, folk. French is more precise in distinguishing between house and family, and the French poetic imagination, therefore, may not be so free as the English to express an apparent attachment to place as such; that is, to the spatial frame alone.

The intent of this cursory semantic analysis has been to expose the uniquely complex significance of home in the English language. I shall turn now to a cursory ethnographic analysis, discussing just two cases: one in which the same unbounded but graded space that is connoted in the English word also appears in the lived experience of a particular group, and another in which the meaning of home varies widely within a single group.

A Note on Ethnography

Jean Jackson has recently studied the marriage networks of the Vaupés, a people of the Amazon basin who live by hunting, fishing, gathering, and swidden cultivation.[6] Each Vaupés settlement consists of a single large structure that houses four to eight nuclear families, and its site is changed periodically. Population density is very low, less than one person per square mile, in a region that is roughly 12,000 square miles in extent. Living along or near rivers in longhouses that are separated by two to ten hours of travel, the Vaupés are frequently travelling and visiting one another. At any one time, Jackson says, "a significant proportion of Indians are not living where they are 'supposed' to."[7] As visitors, they are given the run of the host's land; the men hunt and fish and the women process the manioc growing in the hosts' gardens. The extensive riverine region is multilingual but is well integrated, socially and culturally, through group out-marriage—the rule being that one marries a person who speaks a different language. Most individuals have an extensive knowledge of the Vaupés region. They acquire their information by travel, by living in different sections of the region, and through conversation, committing to memory a store of facts about settlements and topography. The resultant native view is one that recognizes the existence of some spatial clustering of intermarrying language groups, but does not assign them territories and boundaries. Whites and small groups of "foreign" Indians are considered ethnically discrete, but spatially, the author states, "there are no generally recognized natural or artificial boundaries beyond which live people who are definitely enemies or strangers."[8] No strangers "out there"! The notion of "home" for the Vaupés is in some sense infinite, and certainly for these accomplished folk geographers, home—in experience, memory, and myth—has very much the same elastic sense as in the English semantic.

The second case I want to consider has been cited more than once by geographers who assert a phenomenological point of view. Ethnographic accounts of Australian aborigines, such as the Aranda, testify to the great sanctity among them of the ties between people and place. Clan territories are made sacred by ancestors whose spirits continue to live both in place and in their descendants; the stories of the ancestors, woven around local landmarks, tie these enduringly together; these stories are treasured in secret rituals and memorized in symbolic markings on artifacts of stone and wood, called *tjuringas*, that are kept stringently concealed. Of the northern Aranda, the ethnologist, Strehlow, who grew up among them, says: He "clings to his native soil with every fiber of his being . . . love of home, longing for home, these are dominating motives which constantly reappear also in the myths of the totemic ancestors."[9] Elkin goes even further; while clan members can be said to "own" their territory, it may be truer to say that the territory owns them, since "they cannot stay away from it indefinitely and still live."[10]

If we are interested in home from an experiential perspective, we ought perhaps to read a little bit further. These are patrilineal exogamous clans, who take their wives from other clans and give their sisters and daughters to others. When she reaches puberty, a woman leaves her own clan and its territory, ignorant of its secret myths and becoming, as Elkin says, "practically an outsider to them."[11] But women are kept ignorant of the mythic geography of their new location, too. Here is Strehlow again: "No sacred myth ever reaches their ears. Their lips never utter the words of the traditional chants. The ceremonies centering around the lives of their ancestors are carefully hidden from their eyes."[12] Indeed, death is the customary penalty for any woman who might trespass, even though innocently, on the sacred area where the rituals take place.[13] Thus, in the profoundly sacred territory of the clan, for which her menfolk ache in reverent remembrance, Mother—genetrix and nourisher—is forever an alien.

We are then obliged to ask what meaning "home" has for the female half of the Aranda population, since the cultural creation of a mythic attachment to place is meant exclusively for males. The ethnographies are silent. Let us ask the question in a more general way. Is not attachment to what Tuan calls the biological home—the place, that is, of birth and nurture—perhaps always in some degree a male myth?[14] In many societies, female experience is not in accord with a common myth which, at a different scale, gives homeland the name of *patria*, "the land of the *fathers.*" Discordance between male myth and female reality appears in northern India, for example. Entering a village there, one may see, crudely lettered in red, the words: *Grāṁ ka devī mai ki jai!* "Hail to the goddess mother of the village!" If this mythic mother is indeed the embodiment of local, chthonic powers, she is uniquely fitted for the role: every *human* mother in the village comes from someplace else, since the rule of village exogamy prevails.[15] If the myth of home is sex biased, what of the possibility of other biases? Of class, for example? Or of age? Here, I refer to those societies—upper-class Britain, Buddhist Thailand, ancient Sparta—in which the young, usually male, are deliberately "uprooted" in a pedagogy that is precisely intended to wean them from attachment to what is their alleged "biological" home, replacing it with attachment to a mythic one.

Over and over, we find that rather than being the generator of a people, land at one scale or another serves as the chosen symbol of a people's being. Biologic (or quasi-biologic) attachment is to people, cultural attachment is to place. Most ancient regions were named long after their occupants: Graecia (Greece) from the Graeci, a marginal Hellenic tribe; Italia from the Italic peoples. The children of Israel come before the land of Israel. Conversely, the American, who has so often been called "rootless," takes his fundamental identity from his land. But more of that later.

Home and Journey

On one aspect of home noted by Tuan our understanding appears to differ. He speaks of a binary opposition between home and journey and yet does not seem to treat it as a true binary opposition at all.[16] "Journey" for him is an inescapable absence from home, "non-home" if you will. To live, he says, one *has to* take risks in alien places. Journey is *travail.* And this unavoidable journey then defines home: "Home" has no meaning apart from the journey which takes one outside home. It is a firmly home-centered view—I shall call it a "domicentric" view—of human experience. True, he adds elsewhere, "an argument in favor of travel is that it increases awareness, not of exotic places, but of home as a place."[17] He prefaces this statement with a quotation from T. S. Eliot:

We shall not cease from exploration
and the end of all our exploring
Will be to arrive where we started
And know the place for the first time.[18]

Placing this verse in such a context scants the central part of its meaning. The social scientist learns to be wary of poets; and I myself read them, if at all, with the utmost care. What then is Eliot saying in the very first line? "We *shall not cease* from exploration." Here is a true opposition to home, an idea of "anti-home" that recognizes the tension and inner conflict of human existence, individual and social, the tension between order and freedom. It is a recognition that is fundamental to an understanding of human geographic phenomena, one that Carl Sauer captured in the contrast he saw, at a group scale, between peoples of sedentary and mobile bents.[19] People want to make homes and people want to leave home. Home, as Eliot says, "is where one starts from."[20]

The domicentric view is of course the powerful, established one, able to exert enormous social pressure on behavior. Untold psychic damage resulting from contemporary American rootlessness is the thesis of Vance Packard's *A Nation of Strangers*.[21] "Rootlessness" is a stigmatizing term; to be rootless is to be unsound, and, worse, unreliable, unsavory. Listen to Robert Southey (another poet!) in a prose essay written about 1830: "Beware of those who are homeless by choice! You have no hold on a human being whose affections are without a tap-root. The Laws recognize this truth in the privileges which they confer upon freeholders."[22] And he goes on to say: "Vagabond and rogue are convertible terms, and with how much propriety anyone may understand who knows what are the habits of the wandering classes, such as gypsies, tinkers, and potters." More recently in a variation on this theme, Hannah Arendt spoke of "the inherent lawlessness of the uprooted."[23] We know, too, that in the Soviet Union, a code word for Jews in recent years, intended to bring them into disrepute, is that they are "rootless cosmopolitans." "Geopiety," as Yi-fu Tuan has shown, powerfully upholds the stability and order that are centered on a home.

And yet, one is tempted to echo Galileo, mankind does move! From an alternate perspective, to be rooted is the property of vegetables. Set against the myths of home and homeland, we find the myths that challenge them, the myth of the voyager, the adventurer, the mythic quest that takes one forth into and through the world, not reluctantly but eagerly. Let us call these "domifuge questing myths," understanding that they can refer to groups, and, indeed, to whole peoples, as well as to individuals. We know of group myths of this kind that make up part of the fabric of complex civilizations, but they are not unknown among simpler societies. There are intimations of them among the Polynesians, although the most remarkable example of the sort takes us again to South America. There certain peoples of the Tupi-Guarani language group, such as the Apapokuva, have had a tradition recorded by Europeans as early as 1515, of repeatedly moving en masse in a continual quest for the Land without Evil, journeys that have taken them over large stretches of the continent from northeastern Brazil to Peru, and that had not altogether ceased a decade ago.[24] In European folklore, the tale of the Pied Piper of Hamelin appears to be a form of the domifuge quest myth, as seen, however, from the domicentric point of view. (Is it, I wonder, simply coincidence that the place name in the story's title is a cognate of English "home"?)

Several of the great religions of the world have a myth of this kind at their core, urging a transcendence of both society and geography. Here is a contemporary report of one early sect: "They live in their own countries, but only as aliens. They have a share in everything as citizens, and endure everything as foreigners. Every foreign land is their fatherland, and yet for them every fatherland is a foreign land. . . . They busy themselves on earth, but their citizenship is in heaven."[25] These were Christians in the Roman Empire of the third century. A similar dissociation from place and folk is taught by the great Oriental religions, Hinduism and Buddhism, and is frequently realized by individuals. If it is geopiety in China that makes burial in the ancestral ground highly desired, what shall we say of Hindu piety that seeks death far from home, beside a river that will sweep away forever one's ashes after cremation? Is the aspiration to a detachment from place that is thus enacted any less profoundly human?

America itself, the idea of an ever recreated new home—indeed, "a new Heaven and a new Earth"—is yet another version of the myth, taking many elements from the Christian one. Though transformed into a secular myth, it has not been shorn of its religious roots. Eliade says: "The novelty which still fascinates Americans today is a desire with religious underpinnings. In 'novelty' one hopes for a re-naissance, one seeks a new life."[26] How much, then, does this quasi-religious search explain about what we today call "spatial behavior" during the past two centuries of American life?

Doesn't this behavior express the tension between the opposing myths? Can we not recognize this tension in the fact that a century ago when, just as today, one American in five changed residences within a year and one of the most widely treasured household icons was a plaque or sampler carrying the words "Home Sweet Home"? Can we not hear the tension in the theme and tone of our folksongs, with their constant plaintive refrain of "movin' on," although the moving that is foreseen often seems as compulsively ritualistic as the South American wanderings of the Apapokuva?

As a prerequisite to a discussion of the landscape, I have tried to show how the experience and image of home are subject to wide variation both within cultures and across different ones. As exegesis on existing literature, these comments may seem to have lacked balance, since my intention has been to examine the idea of home from novel angles, turning it this way and that, as it were, to display some facets that had been left rather in the shadows. Thus, I have felt it necessary to emphasize the elasticity of the scale of home, both semantically and experientially, as well as the mythic quality of "attachment to place." Since the myth is a cultural artifact that has major importance as an integrator of a group, it is possible to find a people sustained by a myth of a future homeland, as in the case of the Apapokuva, or by that of a once and future home, like the Jews of the Diaspora, or rather, of the Exile, to use the appropriate mythic term. Those Rom ("Gypsies") of Europe who are still nomadic and have suffered for being homeless appear to have only the sustaining notion of a group identity without mythic territorial roots of any kind. They have, of course, territories of circulation, "ranges," as they would be called if the people were Australian aborigines, but no "estates," the myth-invested places hallowed by the spiritual presence of male ancestors. As it happens, the Rom do not have the cult of cemeteries; once a burial is finished, the place of burial is practically forgotten.[27]

How, then, may we apprehend the visible forms, the landscapes, of home? The primary content of home, from what people say, is not material landscape but people. When one is absent, recollection of home is primarily of the human beings there. Without the continuing presence of the sustaining group, the place would no longer be home. It is one's relations with this nurturing and sheltering group as they are associated with the landscape that give it meaning as the landscape of home. At one scale, then, this landscape can be in large part that of the remembered field of familiar experience, within which particular places stand out as the loci of memorable personal events.

Landmarks of Home

At a different level of experience, or at a different scale, the landscape of home may be chiefly a litany of names, pictures, and tales of places that record the direct experience of home by one's people: the members of a family, a larger kin group, or a folk. A family photograph album may have the same role in creating the home of myth as an Australian *tjuringa*, and so may a history book. At yet another scale, when experience of other places suggests that some familiar things at home may be distinctive, these may become generic symbols of home. One becomes then a geographer of the micro-region, if not always a very good one, putting together, perhaps not wittingly, a mental composite of features that tell of home: a profile of hillside, the hue and texture of houses, the pitch of church steeples, the color of cattle. Such home regions will have ragged and changing borders, and certitude may come only with the appearance of the marker of a familiar social field. In Reynolds Price's novel of a Carolina family, *The Surface of Earth*, Rob is being driven home from Virginia, after a long absence, to see his dying grandfather.[28] He is jolted awake in the morning. "At first he was lost; this could be any one of a hundred roads he'd traveled before—narrow, through scrub pine and parched wastes of corn, cotton, weeds, and rough as a cob. Not a human in sight—three starved black cows posted out in an open furnace of a field. . . . A shack swam by and an old man before it, black as any cow they'd passed and dressed to the neck in a snow-white Spanish-American Navy uniform—Deepwater Pritchard, drunk at nine in the morning. Rob looked back and waved. They were twelve miles from home."

These different modes of landscape cognition become mixed together, and may be given very different emphases, in part depending on the scale of home. As the sociologist Robert Nisbet puts it: "Native heath is hardly distinguishable from the human relationships within which landscapes and animals and things become cherished and deeply implanted in one's soul."[29]

One's experience of the landscape of home and one's response to it—how much it is cherished, or whether indeed it is—can hardly be separated from the nature of the human relationships there. Social meaning conditions the lived geography of the childhood home itself: the infant discovering his world by crawling under the grand piano is deterred by his mother from crawling under the kitchen sink. Writing from an authentic black experience of the U.S.A., South and North, Albert Murray asks: "For where else if not the old home place, despite all its prototypical comforts, is the original of all haunted houses and abodes of the booger man?"[30] Having grown up in Alabama, he reminds us that homecoming is "always interwoven with a return to that very old, sometimes forgotten, but ever so easily alerted trouble spot deep inside your innermost being, whoever you are and wherever you are back from."[31]

Our experience of any landscape through the senses is inseparable from the social and psychological context of the experience. It is puzzling that geographers who write in a phenomenological mode often fail to make this point. If to know a place well means to have experienced, for example, the heat of the pavement on an August afternoon, a door-to-door salesman might well experience that sensation differently, and remember it differently, from a college professor on his summer vacation. The overriding meaning of the landscape of home is social. In his book, *Akenfield*, Ronald Blythe has men who had worked on farms as young boys before the First World War tell of the bleakness of their lives in a Sussex village, lives so hard that, grown to manhood in a corner of England that could be its "pastoral heart . . . telling of green days," they left it eagerly at the first opportunity, without ever looking back.[32]

Social creation of a mythic home seeks consciously to play up the uniqueness of place by accenting small distinctions in the landscape, by modifying it idiosyncratically, or by instituting in it a code of local signatures. All these actions imply communication within a larger social group and the existence of conventions making such communication possible. In medieval Europe, the heraldry of lineage—that is, of the social—gave rise in time to the heraldry of corporate institutions that had a fixed territorial base—that is, to a heraldry of place. Heraldic devices appeared for towns and cantons, provinces, and nations; and at a local scale, even wards of cities. Something of this survives in the pageantry and, indeed, in the very meaning of urban life in a few places such as Siena, with its intense, centuries-old, intraurban rivalry, focussed on the annual *palio*. But it needs no formal institution of that kind for the creation of place signatures in the landscape. The carpenter in a Tamil Indian fishing village carefully fashions the prow of a dugout hull, that has itself been imported from the west coast, into a volute that signifies, not ownership by a particular individual, but identity with a particular place.

A consensus that a particular component of the landscape stands for a place—let us use the example of the San Francisco cable car—may develop through a complex interaction between the different views of insider and outsider. The proprietary sentiment and pride that make the feature a

A particular component of the landscape may come to stand for a place. "We ask ourselves whether it would indeed *be* Slovenia without the *Kozolec*." (D.E. Sopher)

symbol of a home place may develop late, and may never acquire, for locals, the affective associations of home. This interplay is more evident in the case of rural landscape features, such as the often elaborately constructed wooden hay-rack, the *kozolec*, of Slovenia in northwestern Yugoslavia. While not completely confined to Slovenia, it is most common there, appearing in many regional and local variations. The poets, folklorists, antiquarians, and geographers of Slovenia have given it great attention in recent years; there is a coffee-table book on the *kozolec* to match coffee-table books on the American barn, and the aspiration to maintain an undiluted Slovenian ethnic identity has elevated the *kozolec* into a self-consciously recognized symbol of the Slovenian homeland.[33] "We ask ourselves," a sociologist in Ljubljana told me, "whether it would indeed *be* Slovenia without the *kozolec*." Yet the emotional association with the *kozolec* expressed at this scale may mean little at the scale of farmstead and village.

Kozolec and cable car are public, consensual symbols of the landscape of home. Private images of such landscapes have not been the particular focus of research on the cognition of the environment, although there is much of value that might be taken from studies showing how knowledge of place is organized, and, a more recent research trend, how that knowledge has been acquired. Let us instead turn again to literature, to find recorded there the human experience of home at various scales, and the shape of remembered images of home that people carry with them. Here I can only offer some impressions, by way of invitation to further thought and reading, from past reading that was, for my present purpose, without design or discipline, but from which I have recovered one or two pertinent passages for closer examination.

A Note on Literature

James Joyce's *Ulysses* ends with Molly Bloom lying half awake, thinking disconnectedly about herself and her past in an unbroken, unpunctuated monologue, forty-five pages long, an imaginative creation that is without parallel in literature.[34] Now and then, memories break in of her childhood and adolescence in Gibraltar, the daughter of an Irish major and his Spanish wife. In these recollections—fragmentary, disjointed, intermittent—Gibraltar, the Rock itself, the view from its high points, the streets, all appear as they were experienced, not as they might be described to another. Molly Bloom is a woman in whom the vital juices are coursing, and she "experiences" with all her pores awake: the sweaty shift sticking to her in summer, the "damn guns bursting and booming all over the shop" on the Queen's birthday, or "the sardines and bream all silver in the fishermens baskets" on Catalan Bay. Places take their meaning for her from the memory of the moments in her life, in her becoming a woman, for which they were the setting. At the same time we can also see how here, as even more elaborately in the case of Bloomsday in Dublin, the action that flows through time and space is expertly modulated by Joyce to have it "make sense" socially in relation to the landscape in which it occurs.

Molly Bloom's flow of memory gives us in sum only fragments of a geography of home. The seasons, the Spanish townspeople, the British military garrison, the heterogeneous Mediterranean communities are all there, but there is almost no focus on a house as such. Home in an understated way was "Gib," and there is little nostalgia: " . . . will I never go back there again all new faces. . ." she interjects at one point. But there is a rush of pleasure at the climax, a memory of bright flowers, pastel houses, and "the queer little streets." In the larger symbolism of the novel, Molly Bloom is

Penelope in an ironical way, but she is also Gea-Tellus, the earth mother of antiquity, no longer a local deity.[35]

Willa Cather's book *The Professor's House* was published in 1925, three years after *Ulysses* had appeared in Paris; but in its view of place and its treatment of landscape it is worlds apart from the earlier novel.[36] The central characters in Cather's book *see* the landscape consciously, they place it in the foreground, and they consciously assign it meaning and value, a value which can be exchanged for others. Thus, for the professor of the title, the chief memory of early childhood is of the water of Lake Michigan, just beyond his family's farm. He remembers "certain human figures against it"—his family, as it happens, out of focus, while the lake is remembered affectionately and in sharp detail forty years later. As an older boy, he is very close to earth, woods, and water, and in retrospect, wise, solitary, observing nature closely and assenting to it. These qualities must be relinquished for marriage, house, career. But an admired student fulfills him vicariously by working in the Southwest, camping in isolation in the Four Corners country, "the kind of place a man would like to stay in forever," he writes. The human habitation in which this young man seems to feel truly at home is the cliffhouse village of the long-dead Indians of the mesa.

What we recognize is that for these two men, the landscape is almost drained of its living human content, and together with it, the social meaning that is the only meaning it has for a Molly Bloom. It appears that in Cather's version of the American quest, closeness to nature is to be balanced by an almost monastic separation from the Joycean world of people; in what some regard as her finest work, *Death Comes for the Archbishop*, these two conditions are combined in her central character.[37] She is not alone as an American writer in seeming to require this separation. One recalls Henry David Thoreau cultivating his place by Walden Pond quite close to town and at the same time propelling himself far away in imaginative space from Concord, seeking company in the thoughts of wiser men than it produces: "Thank Fortune, we are not rooted to the soil, and here is not all the world."[38]

Thoreau's ambivalence toward the society of men may prompt us to ask about social meaning and context in the long tradition of nature-centered poetry, nature being understood here as the negative of *urban* life. Where, in Roman society, should we place the authors of those innumerable eclogues, idylls, and odes to one or another aspect of nature? Horace, showing off the view from his hilltop property to a visitor from the not very distant capital, and extolling his place on the land, might be the very model of a modern exurban gentleman. Horace is perhaps too ironic, too much affectionately aware of *human* nature to fit any neat stereotype. Yet the mythopoeic quality of his work and reputation should not be ignored: in his successors, similar sentiments may become a pose.

Certainly the echoes of this tradition in English Romantic poetry of the early nineteenth century now appear as a mere strained convention. What else can we make of Browning's memory of England as home? Where in April "the lowest boughs and the brushwood sheaf/Round the elmwood bole are in tiny leaf/While the chaffinch sings on the orchard bough."[39] The later echoes of that convention, the celebrated "Letters to the Editor" in the *Times* reporting the first tiny leaf round the elmwood bole are by now what anthropologists call a "cultural performance," like the Green Corn Dance of the Onondaga, a ritualized acting out of a people's sustaining myths, with the actors—in this case, letter-writers and editor—always aware that they *are* acting. Perhaps then we should leave their interpretation to the trained anthropologist.

But even the untrained will have noted that to speak of the myths of a people may not fit this case. What is meant, surely, is a class. We may have suspected this from the absence in Browning's home-thoughts of England, of any shadow of "dark, satanic mills." Evidence of the different class images of England as home appears in one collection of verse, Rudyard Kipling's. They present us with something of a puzzle. Richard Hoggart has described working-class life in Britain in mid-twentieth century as being woven tightly around home and neighborhood,[40] while the education of the officer class, as we know, thrust the boys out at an early age. Kipling himself, born in India, was farmed out at the age of six to spend several wretched years in a foster home in England while his parents returned to India.

Yet in his verse it is the lower ranks, revealed by their dropped initial h's and dropped final g's, who, despite the crude stereotypes they hold of other cultures, nevertheless display some degree of true cosmopolitanism. Think of "Gunga Din," "Fuzzy Wuzzy," "The Mother-Lodge," "Mandalay," or "The Ladies." Complementing this attitude, their view of England ranges from wry tolerance to disenchantment and disparagement. It is a closed, ungrateful society, and compared to the wider world, the landscape of "awful old England" is pallid, unappealing: "blarsted English drizzle" and pale English sunshine, gritty paving stones, beefy shopgirls—and "somethin' gone small with the lot."[41] Kipling was, in Philip Mason's words, "a tribesman without a tribe,"[42] and while he could evoke the immense drama of his Indian home, as he did superbly in *Kim*, he wanted above all to belong to that other mythic home, which he went on to recreate in the poems and stories dealing with a legendary England. When success permitted him to settle in Sussex in a stone farmhouse ("A.D. 1634 over the door"), he exulted to an American friend: "At last I'm one of the gentry."[43] The gentry and their members in the officer caste, unlike the rank and file, could not waver in their sentiments toward England or toward various lesser breeds outside.

One may suspect Kipling of adopting a cockney voice as a literary device in order to express his own ambivalence, but it is likely that he was in fact reporting accurately, since he took pride in his journalist's craft. In his poem "Sussex," he spoke in turn for the gentry and perhaps many others, mythologizing his new home and painting with a loving artist's eye an English rural landscape—line of hill, plant cover, light and shadow, the legend and the mark of history, the forms and the sounds that come from the way man uses the land, and beneath it all, the potent mystical ties to "native heath": "Clay of the pit whence we were wrought/Yearns to its fellow-clay." This is the powerful language of myth, but Kipling's father was a Yorkshireman, his mother was from the Midlands, and the poem is contemporary with the exodus of harried farmworkers from Akenfield.

That exodus is but one instance of a long process of abandonment of rural homes in Europe and the world over. The calculus of the costs and benefits that attend such a move is one that peasants, even in conservative agrarian societies, have long been making. A number of pithy sayings seem to reflect such hard-headed peasant assessments. Latin: *ubi bene ibi patria* ("Where things are good, there's the fatherland"); German: *Es ist überall gut Brot essen* ("It's good to eat bread anywhere"); Hindi: *samat ho to ghar bhalo, nahīn bhalo pardesh* ("When the farming is good, home is fine; when it's not, another country's better").

American Home

Poetry and folk sayings, however, can provide only an uncertain guide to attitudes to home and images of it. Surveys, autobiographical accounts, and field observation should be able to supplement these. A survey during World War II indicates that what was missed by soldiers in the Pacific were, of course, families and sweethearts, and then a variety of minor components of comfortable living or familiar experience: fresh milk, a fresh lettuce and tomato sandwich, the morning paper at the front door, the sound of a train whistle.[44] In my own experience of travel, the things most consistently missed by Americans abroad have been foods of one kind or another (hamburger deprivation seems to be much more severe in India than in Europe) and a variety of articles and facilities suitable for care of the body, thus confirming Horace Miner's observation of the importance of body ritual to the Nacirema (a tribal name derived by spelling "American" backward).[45]

For the black writer Albert Murray, home in Alabama and the rest of the South is recalled by the sounds, smells, and tastes encountered in Harlem that evoke for him a downhome church organ, church meetings, revivals, picnics and camp meetings with white potato salad and sweet potato pies, dasher-turned ice cream after sermon on Sunday—and also roadside beerjoints, Alabama pine-needle breezes, and "blue steel locomotives on northbound tracks."[46] These are, for the most part, recollections of significant community rituals that would also be shared by others. Sound and taste are prominent in all these rememberings. Is this because our store of visual information is so vast that a single cue does little to stimulate the memory, while a single note or odor will produce a clearly remembered sensation that would otherwise be difficult to retrieve? Another conspicuous feature of these recollections is the absence of any reference to landscape in the large, like the "whale-backed downs" and "gnarled thorn" of the poet's vision. Perhaps the questions put do not elicit images of that kind, or the intention to convey some shared national or regional image does not allow of local specificity at the scale of the *pays* or district. Or Yi-fu Tuan is right after all when he says that Americans "lack attachment to place," having a sense of space rather than a sense of place, and that a territory as large as the United States cannot be experienced as a place, except in the abstract.[47]

The claim is not altogether convincing. Instead of demonstrating a lack of attachment to place after the scale has been selected, it would be more useful to let people's sense of place emerge at whatever scale they find convenient. To quote Albert Murray once more, when you get off the subway—Duke Ellington's "A" Train—at 125th Street and St. Nicholas Avenue, "you are also, for better or worse, back among homefolks no matter what part of the old country you come from."[48]

There are good reasons for seeing America as a significant place, and indeed, a home, in people's experience, image, and myth. The landscape artifacts—McDonald's, Mobilgas, drive-in movies, chain stores, Holiday Inns, etc., etc.—that are said to be destroying the distinctive character of American localities, contribute in some measure to the appearance of a common American landscape, thus reducing confusion. Since regional bound-

Artifacts along American highways contribute to the appearance of a common landscape, reducing the immensity of space, giving it coherence and intimacy, allowing us to experience America as a place and, indeed, a home. (P.F. Lewis)

aries are fuzzy at best, so are regional cultural loyalties, promoting eclecticism even in usually secure areas of habit, such as food consumption (tacos stands in Minnesota). The continental flow of highway traffic allows many people to see and experience many places in America and to note their structural commonalities, even if only a small fraction of those who travel do so. George Stewart pointed out long ago that to drive U.S.1 was to see the United States in cross section.[49]

Other media of communication also allow us to know, experience, and imagine America as an integrated whole. Television is surely the most powerful of these. Every week on the network news alone one can see thirty or forty different landscapes, many of them American, and one sees them in color, with sound, and in movement, alive. The explicit content of television may suggest that the country is more homogeneous than it is, but the evidence of regional distinction is there for the watcher to note and to put into proper perspective. On the subject of the news media, one might think of Annie Dillard at Tinker Creek in the Shenandoah Valley, watching the natural life around her, quietly, patiently, through the seasons, reading whatever she can find that might account for what she has observed, yet taking pleasure in the homely cliches of photojournalism that will show from all over the continent, which thus "begins to smack of home," she says, a duck and a cat playing together or a winter scene of a bundled baby bawling in a stationary sled.[50]

Can we deny the powerful integration of vision that a view from the air provides? To fly across the country is to see at one sitting—more than that, to comprehend, to grasp as a whole—the complex interconnections of land, climate, production, and circulation that are woven together in a grand human system. We not only travel about the country easily, we communicate with each other across it, by mail and telephone, frequently. Ties to a particular locality may be frayed, but ties to kin are more robust, and so we find wide-flung networks of kin linking many places rather than being fixed in one. I wonder—idly, since I have no idea—what percentage of the population has at least one close relative or friend living more than, let us say, five hundred miles distant. Would either of them feel uprooted? If we must use botanical metaphors, shouldn't we rather think of rhizomelike connections on the surface of the land forming a dense mat of human interchanges? Here I would add a parenthetical note that is related to the matter of sex bias in the perception of home. Since in many American families today the wife maintains the flow of kin-related information,[51] the nuclear family is usually ambilocal in network space, with a decided tendency to become matrilocal. So perhaps we can say that, freed from the bonds of place, Mother can now be more truly at home.

Is the idea of America really too abstract then for it to be sensed as a place? David Russo has recently urged the following thesis: Americans have always lived simultaneously in various levels of communities; over time, the level of most consequence has shifted by stages from the little community, the town, to the big community, the nation.[52] It would be a mistake, however, to see this as a parallel to European and other Old World experience. Although the level of most consequence may have shifted, Russo is saying that the big community has been in existence for two hundred years. That has given local and regional place a different meaning in America from what it had in Europe. Thus, illiterate pig-keepers in the remotest Kentucky hollows knew themselves to be American a century before the *paysans* of the Cevennes and Gascony had come to think of themselves as Frenchmen. It was Americans who became Ohioans and Iowans and Oklahomans and Oregonians, not the other way around as in Europe. The historic inhibition on displacement arising from identity with locality or region that may linger on in Europe has therefore not had an effect here.

Hence comes the notion of places being exchangeable, which increases as local distinctiveness decreases, and is particularly far advanced in the case of metropolitan areas.

I conclude that America is experienced deeply as a place and as a home. The folk idiom bears this out; it has reduced the immensity of American space, giving it coherence and intimacy. "Out West" and "back East" are the coordinates of this American home, as if one were to say "out front" and "back inside." Yet we could not have had this experience if the rule were not that the stranger be made to feel welcome at home. Peace be upon Robert Frost, but home is not where they have to take you in, it is where they want to take you in. The landmarks of home are the signs that one is welcome. Most of us in academic life know that wherever we may be living, we are to some degree, in the biblical phrase, "strangers in a strange land." Yet the signs in the landscape are there to read, and they can tell us that we are, after all, at home.

Notes

1. Homer, *The Odyssey*, trans. Robert Fitzgerald (Garden City, NY: Doubleday, 1961). Book XIII, lines 194ff, p. 236.
2. Yi-fu Tuan, "Geography, Phenomenology, and the Study of Human Nature," *Canadian Geographer*, 15, no. 3 (1971), 181–92; "Space and Place: Humanistic Perspective," *Progress in Geography* 6 (1974): 211–52; "Place: An Experiential Perspective," *Geographical Review* 65, no. 2 (April 1975): 151–65; "Geopiety: A Theme in Man's Attachment to Nature and to Place," *Geographies of the Mind*, ed. Lowenthal and Martyn Bowden, (New York: Oxford University Press, 1976), pp. 11–39; D. Geoffrey Hayward, "Home as an Environmental and Psychological Concept," *Landscape* 20, no. 1 (October 1975): 2–9; J. Douglas Porteous, "Home: The Territorial Core," *Geographical Review* 66, no. 4 (October 1976): 383–90. Not available at the time of writing was a recently published monograph: E. Relph, *Place and Placelessness* (London: Pion Ltd., 1976).
3. Gaston Bachelard, *La póetique de l'espace*, 2nd ed. (Paris: Presses Universitaires de France, 1958); *The Poetics of Space*, trans. Marie Jolas (Boston: Beacon Press, 1969).
4. Porteous, "Home: The Territorial Core," p. 386.
5. Bachelard, *La poétique de l'espace*, pp. 66–67.
6. Jean Jackson, "Vaupés Marriage: A Network System in the Northwest Amazon," *Regional Analysis*, vol. II: *Social Systems* Carol A. Smith, ed. (New York: Academic Press, 1976), pp. 65–94.
7. Ibid., p. 69.
8. Ibid.

9. T.G.H. Strehlow, _Aranda Traditions_ (Melbourne: Melbourne University Press, 1947), p. 31.
10. A. P. Elkin, _The Australian Aborigines_, 3rd. ed. (Garden City, N.Y.: Doubleday, 1964; orig. pub. 1938, Ref. p. 50.
11. Ibid., p. 92.
12. Strehlow, _Aranda Traditions_, p. 93.
13. A. A. Abbie, _The Original Australians_ (New York: American Elsevier, 1969), p. 138.
14. Tuan, "Geopiety," p. 25; "Geography, Phenomenology and the Study of Human Nature," p. 189.
15. M. J. Libbee and D. E. Sopher, "Marriage Migration in Rural India," _People on the Move: Studies on Internal Migration_, Leszek A. Kosinski and R. Mansell Prothero (London: Methuen, 1975), pp. 347–59.
16. Tuan, "Geography, Phenomenology and the Study of Human Nature," pp. 188–89.
17. Tuan, "Space and Place," p. 235.
18. T. S. Eliot, _Four Quartets_. (New York: Harcourt, Brace 1943), p. 39.
19. Carl O. Sauer, "Sedentary and Mobile Bent in Early Societies," _Social Life of Early Man_, ed. _Sherwood L. Washburn. (New York: Viking Fund Publications in Anthropology), no. 31, 1961, pp. 256–66._
20. Eliot, _Four Quartets_, p. 17.
21. Vance Packard, _A Nation of Strangers_ (New York: David McKay, 1972).
22. Robert Southey, _The Doctor_, chap. 34. Quoted in _Magill's Quotations in Context_, ed. Frank N. Magill, 2nd series, (New York: Harper & Row, 1969), pp. 94–95.
23. "Is America by Nature a Violent Society?" _New York Times Magazine_, 28 April 1968, p. 24.
24. René Ribeiro, "Brazilian Messianic Movements," _Millennial Dreams in Action: Essays in Comparative Study_, ed. Sylvia L. Thrupp (The Hague: Mouton 1962), pp. 55–69.
25. "The So-Called Letter to Diognetus," _The Library of Christian Classics_, vol. I: _Early Christian Fathers_, ed. Eugene R. Fairweather (London: SCM Press, 1953), pp. 205–44, quoted in Clarence J. Glacken, _Traces on the Rhodian Shore_ (Berkeley: University of California Press, 1967), p. 182.
26. Mircea Eliade, "Paradise and Utopia: Mythical Geography and Eschatology," _Daedalus_ (Spring 1965): 260–80; Ref. p. 268.
27. Jean-Paul Clébert, _The Gypsies_, trans. Arthur Duff (Harmondsworth, Middlesex: Penguin Books, 1967). Originally published as _Les tziganes_, 1961. Ref. p. 233.
28. Reynolds Price, _The Surface of Earth_ (New York: Atheneum, 1975), p. 289
29. Robert Nisbet, _Community and Power_ (New York : Oxford University Press, 1964), p. xi.
30. Albert Murray, _South to a Very Old Place_ (New York: McGraw-Hill, 1971), p. 6.
31 Ibid., pp. 5–6
32 Ronald Blythe, _Akenfield: Portrait of an English Village_ (New York: Dell, 1969) pp. 20, 32, 40, 91.
33. Marjan Mušič, _Architecture of the Slovenian Kozolec (Hay Rack)_, trans. J. Golias (Ljubljana: Cankarjeva Založba, 1970).

34. James Joyce, *Ulysses* (New York: The Modern Library, 1934. Molly Bloom's monologue appears on pp. 723–68.

35. Major works of exegesis on the intricate parallels between the novel and Homer's *Odyssey* are Stuart Gilbert, *James Joyce's Ulysses: A Study* (New York: Random House, 1952), and Richard Ellmann, *Ulysses on the Liffey* (New York: Oxford University Press, 1972). The former was done under Joyce's supervision.

36. Willa Cather, *The Professor's House* (New York: Alfred A. Knopf, 1925).

37. Willa Cather, *Death Comes for the Archbishop* (New York: Alfred A. Knopf, 1927).

38. *The Journal of Henry D. Thoreau* vol. I: *1837–1846*, ed. Bradford Torrey and Francis H. Allen (Boston: Houghton Mifflin, 1906), p. 130.

39. Robert Browning, "Home Thoughts From Abroad," (1845).

40. Richard Hoggart, *The Uses of Literacy* (London: Chatto & Windus, 1957), pp. 32–38.

41. Rudyard Kipling, "Chant-Pagan," and "Mandalay," *Collected Verse of Rudyard Kipling* (Garden City, N.Y.: Doubleday, Page & Co., 1925).

42. Philip Mason, *Kipling: The Glass, the Shadow and the Fire* (New York: Harper & Row, 1975), p. 298.

43. Ibid., p. 128.

44. Alfred Schutz, *On Phenomenology and Social Relations* (Chicago: The University of Chicago Press, 1970), p. 296.

45. Horace Miner, "Body Ritual Among the Nacirema," American Anthropologist, 58 (1956): 503–7.

46. Murray, *South to a Very Old Place*, pp. 4–5.

47. Tuan, "Geopiety," p. 28; "American Space, Chinese Place," *Harper's Magazine*, 249, no. 1490 (July 1974): 8.

48. Murray, *South to a Very Old Place*, p. 3.

49. George R. Stewart, *U.S. 40: Cross Section of the United States of America* (Boston: Houghton Mifflin 1953), p. 299.

50. Annie Dillard, *Pilgrim at Tinker Creek* (New York: Bantam Books, 1975), p. 42.

51. I am grateful to Betsy Hansel, who is investigating the spatial structure of American families, for pointing this out to me.

52. David J. Russo, *Families and Communities: A New View of American History* (Nashville: The American Association for State and Local History, 1974), pp. 155–56.

III

American Expressions

The Order of a Landscape

Reason and Religion in Newtonian America

J.B. Jackson

For more than twenty-five years I have been trying to understand and explain that aspect of the environment that I call the landscape. I have written about it, lectured about it, travelled widely to find out about it; and yet I must admit that the concept continues to elude me. Perhaps one reason for this is that I persist in seeing it not as a scenic or ecological entity but as a political or cultural entity, changing in the course of history.

I have come to the point where instead of trying to establish distinctions between landscapes, I try to discover similarities; that is one of the differences between the professional and the amateur traveler: the professional searches for (and finds) differences, and is partial to what might be called a kind of academic romanticism: the establishing of distinct categories. The amateur, on the other hand, is more concerned with finding similarities, with perceiving the universal which presumably lies behind diversity.

That is why the classical or conservative temperament finds the study of history congenial; history repeats itself, establishes patterns and reveals universal laws of human conduct. That, in turn, is why I have become interested in the history of the American landscape, and particularly in one chapter of it, the landscape of the first half of the nineteenth century. I like to think that my interest is more than antiquarian, for to me that landscape represents the last and most grandiose attempt to create an earthly order in harmony with a cosmic order.

The Great Tradition

Let us consider the decades of the mid-eighteenth century, a point in the history of the man-made environment of America when a new and rationalist landscape was about to replace the old medieval, traditional landscape inherited from Europe. It was a time when many relationships in colonial America were changing: relationships between individual members of the family and of society, between individuals and their work, between individuals and their natural environment, and perhaps most significant of all, though hardest to define, a time when the individual was becoming aware of him/herself, and questioning traditional definitions of man. The time was thus approaching when the visible world, especially the man-made world, would begin to reveal those shifts in relationships. For relationships among men and women usually imply spaces, and new relationships produce new ideas of spatial size and location and change or growth. The look of the land, even the look of the house, inevitably reflects those ideas.

We are aware of many of the changes in American society during those decades; every history describes and seeks to account for them: the disintegration of compact and homogeneous towns and villages, the increasing isolation of farms, the distribution and sale of commonlands, and the appearence in the frontier regions of privately financed, speculative settlements; the growth of towns and the decay of some of the older regions; and the growing use of the grid system for land division, not only in towns and cities, but in pioneer communities.

Those are some of the symptoms of change; and the explanations commonly offered for them are no less familiar: the Indian Wars had come to an end on the Atlantic Seaboard, the population was growing, especially in non–Anglo-Saxon elements, and there was an abundance of private capital. All recognize the great importance of these developments, and can see how each in its way affected the Colonial landscape.

We can probably agree on how to interpret them; we can see that they represent a break of traditional ties, social and environmental, and the forming of new relationships based on independence and mobility and rationality, on the release of forces, both destructive and creative, destined in time to alter society and the face of the earth. Relationships, and especially the relationship to the environment, tended to assume a new, more impersonal, more abstract and legalistic form. Land was possessed and exploited not by merely physical means but by contract.

This kind of spatial reorganization is typical of a transitional period (such as we are in now); a period of contradictions and experiments, of the juxtaposition of the old and new, of visual disorder and landscape anarchy. But it is also a period (just as the present is) when there was a search for a new order. And in time that order was discovered.

The New Order

If you ask who discovered it, or when or how, I can give you no precise answer. If you ask what it looked like I can answer that you will see its image on a one-dollar bill, conveniently labelled: "Novus Ordo Seclorum" (The New Order of the Ages). It is a geometrical form, a pyramid, completed by the symbol of an impersonal, all-seeing divinity. However, the date on "The Great Seal of the United States," 1776, does not refer to the origin of the concept: Europe had already formulated the order; it remained for the United States to execute it. Some authorities interpret the image as Masonic, which it may well be, but I think it is best interpreted in terms of Isaac Newton; it was he who gave us a latter-day, classic cosmology, an order based on mathematics and optics.

For such was the order which had begun to transform the European designed environment—the garden, the city, the church and palace, and even the farm—as early as the seventeenth century; an environmental order composed of isolated, independent bodies, moving according to mechanical laws through an absolute undifferentiated space, each one in its prescribed, mathematically measured orbit. Yet in terms of space it was in the New World that the new order first manifested itself, and where it inspired a society based on the predictable and orderly movements of independent, equal individuals, each occupying a portion of the infinite, undifferentiated space made visible in the National Land Survey of 1785.

We have long recognized the role played by Newton's philosophy in the development of modern political theories; it is clear that the rectilinear, rationalist, almost abstract landscape of the early United States was to a large extent inspired by his classic cosmology. But we have paid too little attention I think to the contribution of religion to the new landscape, especially that of evangelical Protestanism; for we cannot really understand that landscape if we omit the emotional ingredient which the Great Awakening provided.

In the first half of the eighteenth century the colonies, and particularly New England, were still loyal to the traditional organization of time and space; space as centripetal and hierarchical, time as a stately procession of inevitable events leading to a dramatic climax. The colonies were, if anything, more conservative in these respects than England; it was in the New World that men believed most firmly in metaphorical gradations of space and of time. It was over here that the seating of the church congregation still indicated social status, that even the graveyard was laid out to indicate rank, and that every procession was arranged according to social position. Time itself, particularly in the life of the individual, was organized into a fixed sequence of events of increasing sanctity.

But this elaborate duplication of ptolemic cosmology was abruptly destroyed (at least in the activities of the Church) by the religious enthusiasm known as the Great Awakening, which began about 1730. There is little need to recall the doctrinal issues involved in the movement, which died down after a decade, only to revive with even greater force in the early years of the nineteenth century. But it accomplished on an emotional and mundane level what Newton accomplished in science and philosophy: it destroyed the old organization of time and the old organization of space. It thereby established a new vernacular landscape.

A brief example will probably suffice. The Great Awakening was largely based on the notion that religious conversion—a radical change in spiritual identity and status—could not only be speeded up, but could be even thought of as instantaneous. Conversion, according to a clergyman writing in 1740, "an absolute, immediate, instantaneous work—darted in upon us like a flash of lightning . . . changing the whole man into a new creature in the twinkling of an eye."[1] This was of course contrary to tradition, for as an orthodox clergyman complained, "conversion is a *progressive* work, and the principles and habits of grace are not infused in us by miracle, all at once . . . [but] acquired by degrees, one virtue added to another, and we grow up to Christian life by insensible gradations."[2]

One result of this reorganizing of the spiritual schedule was the total confusion in the order of the church service itself; no one really knew what was to come next. Changes in the notions of space brought about by the Great Awakening were probably even more important; the movement abolished the idea that there were different kinds of space of varying sanctity. It decreed that all spaces—whether in the church or elsewhere—were of equal value, undifferentiated and even interchangeable. One of the objectional features of the Great Awakening, in the eyes of the conservative element, was the destruction of the notion of territoriality: the idea that the church or congregation was firmly identified with a legally defined, consecrated space: the parish. The first hint of this heresey came when itinerant preachers, invited or uninvited, appeared from outside the parish to conduct the new-style service. Eventually the hierarchical seating arrangement was abandoned, and services were held out-of-doors or in private houses—as well as at odd hours. Finally the new sects built their churches at some distance from the established communities, as if to dissociate themselves from the traditional spatial organization.

Is it unwarranted to see a parallel between this reinterpretation of time and space, and the Newtonian theories? There has recently been published a book entitled *The Religion of Isaac Newton*, by Frank Manuel.[3] Newton wrote much on biblical matters, specifically on the book of Revelation; he was not merely fascinated by symbolism and prophecy, but appears to have been an earnestly religious man. In a passage on the heavenly city he wrote in 1680 as follows: "As fishes in water ascend and descend, move whither they will and rest where they will, so may Angels and Christ and the children of the resurrection do in the air and heavens. 'Tis not the *place* but the *state* which makes heaven and happiness. For God is alike in all places. He is substantially omnipresent, and as much present in the lowest hell as in the highest heaven."[1] The passage can perhaps be interpreted as a mystical version of his views of the cosmos; but what is relevant is the declaration that "it is not place but the state which makes heaven," for it anticipated the doctrine of the Great Awakening and its belief in undifferentiated spaces.

It was, as I have suggested, only toward the end of the eighteenth century that the new order, socially and environmentally speaking, began to assume a concrete form. The Great Awakening, both in its spirit and its date, belongs to the period of preliminary dissatisfaction and search. It is *after* the American Revolution that the vision of the new rational, mathematical order began to inspire the designed environment. And it is hardly necessary to say that the most conspicuous examples of the new style are to be found in the Eastern States: in the layout of Washington, D. C.; Manhat-

It was in the early Midwest that the farmer and villager were first confronted with the new order, but the rational, mathematical aspect of this landscape has been revealed to us only with the coming of commercial flying, as in this view of Illinois from an airplane window. (John A. Jakle)

tan; Baltimore; in such smaller designs as those for the new Burying Ground in New Haven; and above all in the so-called Military Townships of upper New York State.

The Grid

The most imposing example, however, is the grid layout of the Northwest Territory, for it was there in the early Midwest that the farmer and villager and pioneer were first confronted with the new order and obliged to complete its design and refine its crudities. This is the landscape I am interested in, with its peculiar combination of Newtonian or Jeffersonian aftermath, the Great Revival.

It is not an easy landscape to understand. Much of the literature of travel in the Midwest has little value from the landscape point of view. Foreigners who explored the region before the Civil War as far as Missouri or Iowa or Louisiana strove hard and often successfully to entertain, and dwelt on such features of American life as table manners, river steamboats, slavery, political passions, and our tendency to overheat our houses. All followed a more less fixed itinerary: after visiting Washington, New York, and Boston (with a glimpse of Lowell), they went up the Hudson to Niagara Falls, or crossed Pennsylvania to sail down the Ohio, stopping at Cincinnati. Then after a glimpse of "the Prairie," they went down the Mississippi. I have found their set pieces on American scenery all but impossible to read.

I recently studied a number of these works primarily in a search for some early mention of the grid or its visible impact. Aside from a brief passage in Birbeck's "Notes on a Journey in America," written in 1817,[5] I discovered nothing. After describing the township and range system, he says, "All these lines are well defined in the woods, by marks on the trees." That is all. The grid system, despite its present-day visibility, was merely one small aspect of the new American landscape. More impressive and no less common was the isolation of the homes of the settlers, the immense and monotonous distances which separated them, and the absence of sizable towns or villages—space undifferentiated, humanly speaking, from one wooded horizon to the other. The best source of insight into these characteristics are the numerous emigrant handbooks, either in German or English, written throughout the early period. Social historians find them invaluable, for they contain much specific information, simply expressed, on the cost of land, yields of crops, location of markets, distances and means of communication, services and institutions, as well as suggestions as to how to start farming in a new country, how to build a house, what tools and equipment to buy, and so on. Since many of the handbooks were based on personal experience, they contain a good deal of firsthand observation and many shrewd comments on American ways, and though it was not their purpose to emphasize the hardships of pioneer life, they never concealed the problems, economic and psychological, which had confronted them. No matter how much we may have read about frontier existence, these unadorned, essentially optimistic accounts cannot fail to impress us.

Danger in the wilderness was by then no new feature of the American experience; what *was* new was the individual, solitary adjustment to what was still an experimental and incomplete landscape order, a new organization of space that radically affected work and sociability and the business of living. Cabins and houses were hidden from one another by two miles or more of forest; the roads were often impassable and almost invisible. The traditional points of orientation and reassurance—church steeple, tavern, clustering of houses, passersby—did not exist. An Englishman who spent ten lean years in the Midwest peddling Methodist tracts from house to house and who (it is hard to see why) wrote a melancholy but touching account of his travels, incessantly lamented the loneliness of the landscape, the absence of familiar features, of Wednesday night prayer meetings, of cattle grazing in green fields, of old friends. "To be situated in a place where trifling things, as a little yarn for mending, or soap for washing could not be obtained without going a mile or two . . . taking a day's journey to procure what a shilling would purchase in a few minutes in England; to be thus situated was unpleasant."[6]

Movement, the orderly mechanical movement of independent bodies from one space to another in response to predictable laws (usually economic), was characteristic of the Newtonian landscape of the Midwest. Towns grew on a grid pattern in a matter of months, only to vanish in weeks when a railroad line was built elsewhere. The closest neighbor left without warning for Oregon, six months away. "The Americans are a restless people," said the author of one handbook, "And they will go into a neighborhood, work with might and main for a time in opening up a farm; and when they have it in good trim, fences and houses up, land broken and cropped, and an apple and peach orchard planted; sell out, at from 5 to 15 dollars an acre, and settle down again on new land, to repeat the process." Writing in 1850, an English traveler described the American landscape as "ugly and formal"— ugly in its stumps and dead trees, in the litter-strewn yards, the waste everywhere in evidence; formal in its long straight roads or roadways, its large rectangular fields, its bleak, rectangular little houses, its hilltop churches painted a blinding white, its classical placenames, its endless worm fences. One traveler mentioned a worm fence three miles long, absolutely straight.

It is ironic that the rational, mathematical aspect of this landscape has been revealed to us only within the last generation, with the coming of commercial flying. It was there, of course, all the time, but in the abstract, as it were; and the manner in which men and women moved about within it, like counters in a chess game, justifies the cliché of comparing it to a checkerboard—even though, as H. B. Johnson has correctly pointed out, fields were more rectangular rather than square.[7]

The New Individualism

Yet we must not forget the religious ingredient in the early nineteenth century Midwest: its intense individualism and other-worldliness, its concern for the significant instant. For we are dealing with a landscape largely settled and inhabited by men and women, whether Methodists or Baptists or Presbyterians or Mormons, who were participating in and promoting the Great Revival: the landscape of the Methodist circuit rider, the itinerant preacher and peddlar of religious literature, of the revival and camp meeting. A traveler tells of a meeting out on the still vacant prairie with a man who announced that he was the prophet Elijah, and handed out a pamphlet warning of the end of the world. As Eggleston and others have described, it was the landscape of the Millerites, of the Latter Day Saints, of innumerable small millenial sects, all acutely aware of the approaching end of historical time.[8] In the traditional past, the question of meaning of existence had always been "Where?" Where did we belong in the social order? Where was sanctity and salvation? But now the question was When? When was the end to come? When to build a permanent house, when to sell, to move, when to plant and harvest? In the meantime the temporary and makeshift sufficed: houses were good for at least a year or two: burials in the back field or under a tree were good until a later date. A favorite hymn contained the lines:

> Strangers and pilgrims here below,
> This earth, we know, is not our place,
> But hasten through the vale of woe
> And restless to behold Thy face
> Swift to our heavenly country move
> Our everlasting home above.

It was inevitable, I think, that the sense of transiency and mobility affected the manner in which families related to the new environment. Throughout the texts there are occasional vivid landscape descriptions, the result of sharp observation, and written without affectation of sensitivity or romantic emotion; written, that is to say, with detachment: descriptions of a winter dawn over the Indiana forest, of a hot summer day in Iowa spent loading cattle onto a Mississippi ferry, of the Illinois prairie in spring, of an angry rattlesnake. But it would be possible to suppose that the new environment long remained the object of suspicion and even fear; it was a place of unknown diseases, strange weather phenomena. Handbooks made the point that while plants and animals in the Midwest appeared to resemble those of more familiar environments, they were really very different. An established literary tradition assures us that America developed a new feeling for nature in the early nineteenth century.[9] I suspect that this was confined to a small class, and was moreover a European importation of little widespread acceptance. To read the descriptions of the pioneer houses and ways of life is to discern a very definite urge to exclude the untamed outside world. The pride of the homesteader was not in his wider experience of nature but in the cleared field, the brussels carpet on the floor, the store-bought windows which kept out fresh air, the daughter who had taken piano lessons. In the course of not many years the fertility of the soil decreased, and yet manure accumulated so deep in the barn that the only solution was to build a new one.

These traits have been analyzed and criticised by writers and historians, as evidence of American rapacity or ignorance, or (in loftier terms) as forerunners of capitalist exploitation of natural resources. Perhaps those are sufficient explanations. But one cannot help feeling that the spirit of the Midwestern landscape and of the society which created it was not opportunism and greed but unquestioning acceptance of the authority of a revealed truth: the truth of the Bible, the truth of mathematical formulas and rational philosophy. The more one reads about the individualistic behavior of the early settlers the more one is struck by the fact that unlike the contemporary individualist they were not "doing their own thing," as the phrase goes, but acting on what to them has some unimpeachable authority: the Bible, the Constitution, the writings of Thomas Paine or Locke or perhaps even Newton. It was this reliance on the authority of the Book, sacred or profane, that made it unnecessary in their eyes to adjust to the immediate environment in any very effective way. Truth has been revealed in the cosmological order, they believed: why try to find it elsewhere?

A group of cattle-dealers spent the evening in a tavern in Wisconsin reviling the Mormons for their distortion of the Bible; the men cited chapter and verse to prove the heresies of the Mormon faith. A Swiss teacher, exploring the region for a likely place to settle, had the temerity to reproach them for their intolerance. In what he later described as Baroque English (which caused his audience to laugh), he reminded them that God's truth was revealed in two ways: in Scripture, and in the aspect of the material world around us. He said that both the Mormons—this was in 1853—and other Americans, including his hearers, were wrong to rely too much on the Book. He saw this as a great weakness of religion in America and prophesied the coming of a new period, a new order, when science would reveal God's truth in the visible world.[10]

He was right in his prophesy: after the Civil War we discern a new, more pragmatic, more scientific attitude toward the environment and its exploitation. And when that became general we can say that the classical

Newtonian cosmology had been finally rejected and that the classical American landscape was doomed. A new cosmology was even then beginning to take form.

Notes

1. Alexander Garden, "Regeneration, and the Testimony of the Spirit. Being the Substance of Two Sermons," quoted in *The Great Awakening*, ed. Alan Heimert and Perry Miller (New York: American Heritage, 1967), p. 57.
2. Samuel Quincy, "The Nature and Necessity of Regeneration," quoted in Heimert and Miller, *The Great Awakening*, p. 488.
3. Frank E. Manuel, *The Religion of Isaac Newton* (New York: Oxford University Press, 1974).
4. *Op. cit.*, p. 101.
5. Morris Birkbeck, *Notes on a Journey in America* (London, 1818), p. 71.
6. John Eyre, *Travels in America* (New York, 1851), p. 63.
7. Hildegard Binder Johnson, *Order Upon the Land, The U.S. Rectangular Land Survey and the Upper Mississippi Country* (New York: Oxford University Press, 1976).
8. Bernard De Voto, *The Year of Decision 1846* (Boston: Little Brown, 1943).
9. Hans Huth, *Nature and the American: Three Centuries of Changing Attitudes* (Berkeley: University of California Press, 1957).
10. Heinrich Busshard, *Anschauugen und Erfahrungen in Nord Amerika* (Zurich: 1953), p. 372.

Symbolic Landscapes

Some Idealizations of American Communities

D.W. Meinig

Three Landscapes

Every mature nation has its symbolic landscapes. They are part of the iconography of nationhood, part of the shared set of ideas and memories and feelings which bind a people together.[1]

The topic is a complex one, fraught with nuances and different expressions at various levels of social consciousness, but the existence of the phenomenon seems clear. One need not argue for some mystical bond of Blut and Boden, one need only point to the kinds of landscape images widely employed because they are assumed to convey certain meanings. The simplest examples are those which are clearly identified with specific major institutions or events, such as, in the case of the United States, the White House and Independence Hall. In the great majority of cases the pictures of these buildings put before us in various media are meant to evoke responses which have little to do with the appreciation fo their specific architecture as buildings; rather they are assumed to prompt some connection with our national institutions and history.

There are also landscape depictions which may be powerfully evocative because they are understood as being a particular kind of place rather than a precise building or locality. Among the most famous in America is the scene of a village embowered in great elms and maples, its location marked by a slender steeple rising gracefully above a white wooden church which faces on a village green around which are arrayed large white clapboard houses which, like the church, show a simple elegance in form and trim. These few phrases are sufficient to conjure an instant mental image of a special kind of place in a very famous region. As the author of a recent guidebook confidently stated:

> To the entire world, a steepled church, set in its frame of white wooden houses around a manicured common, remains a scene which says "New England."[2]

Our interest is not simply in the fact that such a scene "says" New England, but more especially in what New England "says" to us through the medium of its villages. That it says something which is widely appreciated seems clear from the nationwide popularity of such scenes: on calendars, patriotic posters, Christmas cards and other religious materials; in the many prints and paintings which adorn the walls of homes and offices and public places; in their use in advertising, especially for products or services related to the family, home, and security. Just what meaning is intended and what is received from such depictions might be difficult to know with any precision, but drawing simply upon one's experience as an American (which is, after all, an appropriate way to judge a national symbol) it seems clear that such scenes carry connotations of continuity (of not just something important in our past, but a visible bond between past and present), of stability, quiet prosperity, cohesion and intimacy. Taken as a whole, the image of the New England village is widely assumed to symbolize for many people the best we have known of an intimate, family-centered, Godfearing, morally conscious, industrious, thrifty, democratic *community*.

That is of course a projection from an actual landscape and society. The New England village was a distinctive American creation of a very distinctive society. Although many of its features such as the concept of town and village, the arable strips and common lands, and the centrality of the church obviously had English antecedents there were significant differences in form, content, and function. In America the Puritans, who had been

A steepled church set in a frame of white houses around a common says "New England," and is among the most powerful items in the iconography of America. Lyndon Center, Vermont (Vermont Development Agency)

no more than loosely associated fragments within the larger society of England, formed a relatively homogeneous group which attempted at least in some places to create "Christian, utopian, closed, corporate" communities.[3] Their early settlements stamped a distinct imprint upon the glaciated lowlands of New England, and in some loosened form that pattern was spread over nearly all of New England, most of Upstate New York and northernmost Pennsylvania, major districts of Ohio, Michigan, and Indiana, and sporadically over a broad expanse of the Upper Middle West.[4]

The New England village as a landscape form was thus evident in some degree well beyond its source region, and its fame as a distinct kind of community and setting spread far beyond any local imprint. It became national, and the means of such effective diffusion seems generally evident. Throughout the nineteenth century New Englanders dominated the writing of American history and literature, they were the most powerful influence upon American education, and they were self-appointed guardians of American moral authority. It seems reasonable to assert that in association with such activities an idealized image of the New England village became so powerfully impressed upon such a broad readership as to become a national symbol, a model setting for the American community.[5]

There are other model landscapes of American community, emerging out of our national experience with other regions and other times. Certainly a major successor and rival to the New England village is a scene focused not upon the church and village green but upon a street, lined with three or four-story red brick business blocks, whose rather ornate fenestrations and cornices reveal their nineteenth century origins. Above the storefronts and awnings are the offices of lawyers, doctors, and dentists, and above these the meeting rooms of the various fraternal orders. A courthouse, set apart on its own block, may be visible, but it is not an essential element, for the great classical columns fronting the stone temple of business proclaim the bank as the real seat of authority. This is Main Street, and parallel with it lies Church Street, not of *the* church, but of church*es*: Methodist, Presbyterian, Baptist, Episcopalian, and if there are Yankees present, Congregational. Close by is the academy and perhaps a small denominational college. The residential area begins with big Italianate and Victorian houses on spacious tree-shaded lots and grades out to lesser but still comfortable homes. On the other side of town, below the depot, are the warehouses and small factories. And around it all lies a prosperous farming country dotted with handsome farmhouses and big red barns.

We may well refer to this landscape as Main Street of Middle America. The basic order is linear: Main Street running east and west, a business thoroughfare aligned with the axis of national development. It is "middle" in many connotations: in location—between the frontier to the west and the cosmopolitan seaports to the east; in economy—a commercial center surrounded by agriculture and augmented by local industry to form a balanced diversity; in social class and structure—with no great extremes of wealth or poverty, with social gradations but no rigid layers, a genuine community but not tightly cohesive; in size—not so small as to be stultifying nor so large as to forfeit friendship and familiarity. In this generalized image Main Street is the seat of a business culture of property-minded, law-abiding citizens devoted to "free enterprise" and "social morality," a community of sober, sensible, practical people. The Chamber of Commerce and the Protestant churches are naturally linked in support of "progress" and "improvement." For many people over many decades of our national life this is the landscape of "small town virtues," the "backbone of America," the "real America."

And of course this, too, is an idealized version of an actual landscape, one which emerged in the Ohio country, expanded broadly over the Middle

The classical columns fronting the stone temple of business proclaimed the bank as the real seat of authority, as in this rather subdued version of main street in Bath, New York, just to the east of the archetypical Ohio Country. (Milo Stewart, New York State Council on the Arts)

West, and reappeared in some degree in parts of Colorado, the Sacramento Valley, and the Great Columbia Plain. As a cultural form it drew upon three regional societies of the colonial seaboard. The clearest antecedant is southeastern Pennsylvania with its diverse amalgam of peoples forming social neighborhoods in the rolling productive farmlands and giving rise to thriving market towns.[6] But Ohio was not a conscious imitation and it easily blended in elements from New England, especially in reference to religion, education, morality, and Virginia, especially in political life and forms. Arising in our "national hearth," this experimental ground of the new republic during the first half-century of our national life and characteristic of such a broad realm of our most productive lands where so many millions found some real substance to the American Dream, it is not surprising that this Main Street became an enormously influential landscape symbol, widely assumed to represent the most "typically American."[7]

Created during the canal and early railroad age of the mid-nineteenth century, such landscapes were readily adapted to accomodate the electric interurbans and street cars of the turn of the century. Early automobiles were also quickly and proudly incorporated, but in time the automobile proved much too powerful to be contained and domesticated within such a landscape. It was such a revolutionary instrument, so penetrating and pervasive in its impact upon American society, that it created its own landscape, its own physical and social form of community.

And so let us visualize a third scene: of low, wide-spreading, single-story houses standing on broad lots fronted by open, perfect green lawns; the most prominent feature of the house is the two-car garage opening onto a broad driveway, connecting to the broad curving street (with no side-walks, for pedestrians are unknown and unwanted) which leads to the great freeways on which these affluent nuclear families can be carried swiftly and effortlessly in air-conditioned comfort to surfing or skiing, golfing, boating, or country-clubbing, as well as to the great shopping plazas and to drive-in facilities catering to every need and whim.

This is suburbia, but more specifically it is California Suburbia. Of course suburbs did not originate in California. In American they began to appear on the outskirts of many cities in the latter nineteenth century. Sam Bass Warner's well-known book *Streetcar Suburbs* refers to a major example of a type and era.[8] But the idealized suburban landscape awaited the development of something more: not suburbs as mere adjuncts of older urban areas, but a discrete and independent landscape, detaching the term from its literal meaning: not *sub*-ordinate, but the dominant pattern. That awaited the mass-ownership of automobiles, giving every family autonomous, discretionary mobility over wide areas, which, in turn, allowed the development of entirely new "urban" areas designed for the automobile. It seems quite clear that the major culture hearth of this development was Southern California.

A streetcar suburbia of great attractiveness had been developed in Southern California in the wake of the great boom of the 1880s. Influenced by the earlier promotion of subtropical agricultural colonies it had a marked horticultural look with homes surrounded by an effusion of flowers, gardens, and groves. The small irrigated plot amidst the orange groves had proved a powerful attraction to Midwestern migrants, but the emphasis steadily shifted from practical agriculture to simply the enjoyment of life in a wonderfully pleasant environment. It was as an extension from this distinctive kind of regional settlement that the new landscape of Automobile Culture rapidly emerged in the 1920s.[9]

Houses amidst an effusion of flowers, gardens, and groves between the mountains and the sea in the sunny subtropics of Southern California was a new suburbia of great attractiveness. (Los Angeles Public Library)

As Frank Donovan has stated, "the greatest single factor" in the transformation of America in the twentieth century

> was a new concept of the role of the automobile. Starting as a rich man's toy or the plaything of sports, it had become, during the teens, a dependable utilitarian means of transportation, accepted by farmers and the middle class. Now it suddenly became a way of life for all Americans.[10]

Our concern is not with the development of the automobile but of the landscape developed as a result of it. A wide array of new elements was involved, such as new types of houses incorporating garages and carports, new street patterns and road designs, new kinds of automobile service stations and drive-in facilities, motels and shopping plazas, auto clubs and free road maps. Many of these items were developed elsewhere, but taken together as a new culture complex shaping its own landscape it appears first and most thoroughly in Southern California. The East built the cars, but California taught us how to live with them.

Southern California was a strong growth region in the 1920s. Extensive areas of open land were being urbanized and thus designers could create a new landscape to fit the automobile rather than adapt older forms to accomodate a radical innovation.[11] But there was more to it than just room for expansion. (Florida was booming too, but had little noticeable impact on the national landscape.) Southern California was also giving birth to a new kind of society. It might be characterized as a leisure society, not because most of the people were so rich they need not work, but because it was based on a very different attitude toward work which made leisure a positive good, a definite break with older Puritan and Middle American mores. Southern California was the chief source-region of a new American life-style which has been expanding and elaborating for more than fifty years, featuring a relaxed enjoyment of each day in casual indoor-outdoor living, with an accent upon individual gratification, physical health, and pleasant exercise. It was a style which took maximum advantage of a distinctive geographic setting. The patio, swimming pool, and backyard barbeque, furniture and clothing designed for relaxed daily living, the enjoyment of sun bathing, swimming, surfing and tennis were all beautifully appropriate to the sunny summer-dry subtropics amidst the orange groves and flowering shrubs between the mountains and the seashore. The automobile was an integral and essential part of this new individualistic, informal, immediate life-style. It was an assumed feature of major importance to the design of clothing, houses, services, whole cities, and what we have termed the landscape of California Suburbia was the general result.

Furthermore, although distinctively regional in some of its basic elements, this image was idealized and rapidly diffused to the nation. It was a powerful image, for it combined a very attractive physical landscape designed to serve a very attractive new way of life; it was associated with a region which had a mythical quality about it as part of the persistent deep psychological drives of the westward movement;[12] and its depiction was carried to the American public by an unprecedentedly powerful propaganda medium: the cinema. The emergence of this new life-style and of automobile culture was intimately and complexly associated with the emergence of the movie industry. All were bathed in the same glamour, the same association with that which was regarded as "modern" and fashionable. Thus Hollywood, mostly unconsciously, perhaps, put before the eyes of the world a selective, idealized California landscape as if it were the best in American life, an obvious standard to strive for, a model for the future. And for half a century the nation was so remade in imitation that the stamp of California can be seen on the American domestic landscape even at the farthest environmental remove from that summer-dry subtropical hearth. So, too, the generalized concept of suburbia became "the equivalent of small town America as the symbol of the country's grass roots and the fountainhead of the American Way of Life."[13]

There are other examples of such symbolic landscapes in America. The Southern Plantation is a famous one, but it is generally understood to be regionally limited even though it has had some impact well beyond the Old South. But I believe that the three I have briefly noted, the New England Village, Main Street of Middle America, and California Suburbia, are the three which have been most influential at the national level with reference to idealized communities for family life. Each is based upon an actual landscape of a particular region. Each is an image derived from our national experience which has been simplified, beautified, and widely advertised so as to become a commonly understood symbol. Each has in some considerable degree influenced the shaping of the American scene over broader areas.

If I am at all correct in these assertions (and they are no more than that at this stage, the product of reflection rather than focused research), there remains a lot of interesting work to be done to refine, assess, and apply this basic idea in the interpretation of American culture.

Six Questions

The general topic of symbolic landscapes impinges upon the very essence of what we mean by "American," of what we understand to be the nature of our society and its essential history, and thus ramifies far beyond the competence of any one inquirer or even any particular discipline. The topic would seem to be central to "American Studies," that inherently interdisciplinary enterprise, and because it focuses directly upon relationships between a culture and the landscape it creates, and upon a few specific regions and their influence upon the nation, it would seem to be a topic particularly appropriate to cultural geographers. What follows is a suggestion of some ways to explore this complex terrain, addressed especially to geographers but with implicit recognition that really successful exploring parties would have to include a variety of specialists from several disciplines.

These suggestions are set forth in a sequence of questions which lead, in general, from source to symbol, and from the past to the present and future.

We might well begin our inquiry by asking:

WHAT WERE THE LANDSCAPES WHICH HAVE SERVED AS THE BASES FOR THESE SYMBOLS REALLY LIKE? We may feel we know them well, but perhaps we have been deluded by the very power of the symbols. When we attempt to penetrate the familiar generalizations and clichés about the New England Village, Main Street, and Suburbia, we may be startled at how narrow and uneven are the foundations upon which these stereotypes rest.

Certainly the geographical literature is very thin. We have produced some useful materials, but have hardly begun to put these together to build a much more comprehensive understanding of the making of the American landscape. We need to know not only more about the physical and social character of these three kinds of communities, but much more about how these varied region by region and have been altered era by era. As Zelinsky has recently noted:

> We know surprisingly little about the forms and appearance of the vast majority of the cities and towns of North America. We know even less about the meaning of these phenomena in the cultural scheme of things.[14]

This task is an inherent responsibility of geographers, for other fields simply do not view the topic in such spatial terms. For example, those intensive studies of several New England villages and towns by social historians tell us much we need to know about those places (even if they do not always detail the landscape in ways we might wish), but how representative such places were of how large an area remains quite uncertain.[15] A basic step would be a map of morphological types in New England so as to begin to build a geographic context for the assessment of individual cases.[16] Not all, perhaps not even most, New England colonists settled in clustered villages, and the transformation of Puritans into Yankees must surely have been discernible in the landscape of settlement; but we do not as yet have an adequate description of the processes and results.[17] And, of course, we know that over much of southern New England and more selectively elsewhere industrialization and later immigrations so extensively transformed these Yankee communities as to obscure or almost efface those elements symbolic of the region.

   Thornton Wilder's classic *Our Town*, one of the most popular plays in the history of the American theater, was set in New England at a very specific time and place. The playwright assumed it was a landscape so familiar to audiences anywhere that he relied not on stage scenery but on a few descriptive phrases to trigger their imaginations: As the curtain rises the Stage Manager walks onto the blank stage and proceeds to tell the audience about the setting for the play. He says that the name of the town is Grover's Corners, New Hampshire, just across the Massachusetts line and that the first act will show what happens on May 7, 1901. He then proceeds to describe in words and gestures "how our town lies," indicating the line of Main Street and of the railroad, pointing to Polish Town across the tracks, mentioning in an aside that some Canuck families live there. He then notes the locations along Main Street of the Town Hall and Post Office, the Congregational, Presbyterian, Methodist, and Unitarian churches. The Baptist Church is down by the river, the Catholic over beyond the tracks.[18]

   If this sounds more like the landscape of Main Street than of the New England Village, it is because such a village had gotten caught up in the booming business culture of the nineteenth century and been greatly enlarged and diversified. But the special New England elements are there, not only in later references to 1670s gravestones in the town cemetery and to the blanket factory, but in the presence of those "Canucks" (French Canadians, presumably lured by that factory) and the prominence of the Congregational and Unitarian churches. The question here for geographers to investigate is how representative *Our Town* was of the actual landscapes of that time along the Massachusetts–New Hampshire border. And, if there was a close similarity, to assess how representative the communities of that border zone were of New England and of Yankee-influenced regions. The same kinds of questions and need for geographical investigations apply to the assessment of the actual landscapes underlying Main Street and Suburbia.[19]

   Thornton Wilder's play was a conscious attempt to create an idealization of community life,[20] and that fact can serve to pose a second question:

HOW DO ACTUAL LANDSCAPES BECOME SYMBOLIC LANDSCAPES? Such a query actually contains two rather distinct lines of inquiry: a) What are the means of selection of particular kinds of localities for idealization? and b) How do these chosen scenes become generalized and imprinted in the public mind? I shall touch on the first of these later in the context of another question. As for the second, we might begin with inventories of landscape depictions in all kinds of literature and other visual media. Occasionally geographers have made some assessment of important fictional accounts in relation to the actual landscapes upon which they were based.[21] We could well do with many more, but in terms of this question we should probably give emphasis to a wide body of more truly "popular" materials: magazines, newspapers, and advertisements; comic books and textbooks; calendars and greeting cards; photographs, paintings, and sketches, posters and wallpapers. Obviously this is an enormous bulk of materials and any survey will have to be selective. For the purposes of the theme of this essay, special attention might be given to those landscape depictions which are overtly propagandistic in relating to the "American way of life," to settings which are assumed to conjure some association with basic values and mores of idealized domestic life.[22]

For the past sixty years the cinema has been widely assumed to have had a powerful impact on popular attitudes toward many things. It has displayed an enormous range of landscapes to millions of people, and within those myriad scenes there have been some which were obviously meant to convey settings representative of some concept of the ordinary good and happy life in America. An efficient beginning for an investigation of these would be a study of the character of the outdoor sets which the major motion-picture companies maintained on their lots during the peak of the Hollywood era circa 1920s–1950s. One suspects, for example, that "small town America" was filmed time and again on essentially the same set in which the facades of an idealized "typical" Main Street, church, and a few residences had been created. A logical extension of such an inquiry would be an inventory of the actual towns which were used for on-location filming of similar kinds of shows.

Such an inventory of types of scenes is of course only a beginning, but it would allow us to make some ready inferences. If we have some understanding of the intent behind the use of a particular type of scene we may assume that the user believes that such landscapes will indeed provoke the proper response. For example, specialists in advertising could presumably provide us with information, supported in some degree by research, on their understandings of the common psychological associations which various kinds of scenes suggest.

These suggested studies only deal with the basic evidence of use and the assumptions of those who employ landscape depictions for some specific purpose. They will not get us very far in understanding *why* people regard these landscapes as symbolic of certain values and ideals, but let us turn next to another more limited question:

HOW CAN WE ASSESS THE IMPACT, THE POWER, OF THE SYMBOL? Most social scientists would likely start with people and ask them questions which might reveal their attitudes about such matters, but I think geographers might best start with the landscape itself to see what we can find there of how substance is shaped by the symbolic. That allows us to deal with results rather than opinions, with the past as well as the present, and is the logical point of departure for a field which is fundamentally concerned with environments and places.

Take a simple example: those crude little steeples of wood, or metal sheets, or, nowadays, plastic which one so often sees perched on little meeting houses of various pentecostal sects all over America. Obviously no architect was responsible for them. They surely represent a widespread folk idea of what a church *should* look like. And where does that idea come from? Perhaps from the local examples of the large churches of major denominations, but I suspect a more influential model is something closer in size and materials and assumed to be closer in social concept: the small, white, wooden church of our God-fearing ancestors—and that image is almost pure New England.

Or consider a more extensive phenomena: the diffusion over the entire nation of a succession of California housetypes, the bungalow, mission stucco, ranchhouse, and various "contemporary" styles, strongly affecting the character of the largest growing sector of our metropolitan landscapes. There were always, of course, competing styles of what a nice, middle-class modern house should look like. In the Northeast, Cape Cods and various pseudo-colonials have long competed effectively with Californian models. We need studies in historical geography which will sort out and reveal regional patterns of the presence, proportions, and timing of these distinctive styles.

Or take a more subtle example: consider how New England villages have been remodeled and tidied up to fit the symbol. A very selective migration has moved into a great many of these villages, armed with considerable wealth, taste, and a vision of what a New England Village *should* look like, and has proceeded to dress up those villages and to build houses and shops to conform to those calendar and Christmas card depictions. Presumably few of these people would want to try to recreate the Puritan or Yankee life-style, yet there remains a power in that landscape as a symbol of an attractive scale and type of local society, and some people do indeed move to such villages to try and recover some more intimate sense of community.

Such topics are small parts of the broad study of the making of the American landscape, a task which despite a few really creative contributors remains in a very early stage and will require the development of an array of new kinds of literature.[23] When we have a much better grasp of this kind of history we will be in a better position to address another question:

HOW DO WE DEFINE AND ASSESS THE SIGNIFICANCE OF THE DIFFERENCE BE-TWEEN THE IDEAL AND THE REAL? That there is a difference is of course inher-ent in the very idea of a symbol, but it seems important to take a close look at the nature and scale of that difference.

Take those lovely New England villages just mentioned. The co-existence of two "realities," one consciously tied to the symbolic, the other oblivious of or perhaps antagonistic to such an image, can be a vivid and at times bitter dimension of the current scene. For example, the *New York Times* recently reported on a "classic New England conflict" in Walpole, New Hampshire, over a proposal to build a large pulpmill on the outskirts of the village.[24] During a day-long hearing, "the area's more well-to-do, sophisticated residents . . . wearing goose down parkas and tweed coats" fought the mill as a desecration, while local businessmen and workers "in rough work clothes and boots," whose families had been natives there for generations favored the mill as a source of money and jobs. The reporter noted that the division was seen as an expression of class distinctions which are vivid in the landscape:

> This is, in effect, two towns [villages] within the same border. Walpole itself is a picture-postcard New England town with white Greek revival homes flank-ing the main street in the town common. . . . Long the site of summer homes for the well-to-do, it has been attracting retired people from other areas, and the tiny town center supports an art gallery and a gift shop.
>
> To the north on Route 12, lies the sharply contrasting hamlet of North Walpole . . . where drab and sometimes ramshackle frame houses huddle closely together by the railroad tracks.

The one group is living consciously and determinedly in a symbolic landscape, having selected that setting for a special way of life, one widely understood and admired by Americans. For such exurbanites, the New Eng-land Village is a way of connecting their lives to an idealized past.[25] The other group continues to face the harsh realities of how to make a living out of this hard New England ground, a problem which has been driving such persons away from these New England villages for generations, and a problem which is simply ignored in the idealizations.

Or consider Main Street. In the idealized version, the people of such communities are usually considered to be very largely middle-class white Anglo-Saxon Protestants. In some areas such towns were made up of such populations, but a panorama of the landscape of most towns in the Middle West, and especially those which were best representative of latter-nineteenth century business culture, would reveal several other groups: Irish Catholic laborers down along the canal, who were regarded by the spokesmen for Main Street as a dirt-poor, boozing and brawling bunch led by papal agents (the Catholic Church was not on Church Street); Poor Whites on the other side of the tracks, regarded as drifters and dregs who were illiterate, unskilled, and unsuccessful through their own fault; and in the shantytown down along the riverbank, the "colored folk," simple, shiftless, irresponsible, free but never to be equal, destined to be hewers of wood and drawers of water; and, near the end of the century, new social districts crowded alongside the new industries, full of "foreigners" of strange tongues and clothes and manners, huddled around their own churches, taverns, and social clubs.

Such people and their habitations and facilities show up marginally if at all in the symbolic landscape. They are not really welcome on Main Street, they certainly are not part of the idealized community which was considered representative of basic American virtues. Thus the symbol did not encompass the actual diversity of its landscape reference; and the gap between the symbolic and the real, in terms of kinds of people and the way the social system actually operated, came to represent such a distortion that Main Street became widely discredited as a community form, as a large body of American literature attests. Much of the critique focused on size, intimacy, and the emphasis upon business, "progress," and Protestant morality—those very features emphasized in the idealization—as stultifying limitations. When Edith Wharton said in 1927 that "The Great American Novel must always be about Main Street, geographically, socially, intellectually," it was at once an expression of her urbane contempt (and probably jealousy of Sinclair Lewis) as well as a testament to the power of the symbol.[26]

Despite an avalanche of denunciations in novels and plays and chronic derision in jokes and urban folklore, the small town has retained a powerful claim on American sentiment.[27] The meticulously constructed

Main Street remains the most popular of the many sections of the massively popular Disneyland. Its "two blocks of miniaturized architecture," modeled on a Missouri prototype, is a tangible symbolic landscape, a focus for the persisting nostalgia for what is imagined to have been a better scale and form of community life than most of its visitors now enjoy.[28] When Americans dream of the ideal place in which to live, the concept of "small is beautiful" has been a powerfully persisting counterpoint to the general national obsession with growth and bigness.

The case of Suburbia has some parallels with that of Main Street. There is a large literature telling us how tarnished a symbol it has become. Nearly twenty years ago David Reisman summarized the view of most social commentators of the time by saying they had come to regard suburbs with

> more loathing than love, finding them homogeneous, conformist, adjustment-oriented, conservative, dull, child-centered, female-dominated, anti-individualist—in a word, impossible places to live.[29]

That certainly suggests the emergence of a wide gap between the symbol and the substance. In part there was also a gap between two actual kinds of suburban landscapes: between the citrus grove suburbia of Southern California of the 1920s (the primary basis of the symbol) and the Levittowns and other massive suburban creations in Megalopolis of the 1950s, (the prime focus of sociological study). And that reinforces the main point of comment on this question: that there is an important task for historical geographers in defining what the landscapes underlying the symbols and the regional variations in those basic types were actually like.

Although all three of these symbolic landscapes still exert some power in American society, it is certain that all three are diminished in influence and, more significant, none of the three is soundly based upon the actualities of community life today. Thus none can be regarded as a satisfying image for the future. Before considering directly the implications of that, however, it will be useful to reflect further upon these three, taken together, and ask:

WHAT DOES THIS THREEFOLD SET OF SYMBOLS TELL US ABOUT AMERICA? Geographers might first note that it is a *set of regions*, with the clear implication that New England, the Middle West, and Southern California have been successively critical areas in our developing national cul-

Converting the symbol into tangible form: Walt Disney's masterpiece of contrived fantasy. (© Walt Disney Productions)

ture. Each in turn has seemed to embody the best, or the essence, of America, a model for the nation. Is that a correct assumption of influence? If so, why did these particular landscapes so serve? Was each in its time the most creative region, the most vibrant powerful center of American development? Such an interpretation would not be exactly concordant with common understandings of the American Core Area, of where the major centers of power were through much of our history. Perhaps we need a reexamination of those common understandings.

Or is it a case not so much of power in some manipulative sense as of seeing our experiences in this succession of regions as the most important we have had as a national people? That, too, seems hard to square with common interpretations: it simply leaves out far too much of our history. Perhaps each region has been critically related to some larger American self-image, representative of what Americans wanted to believe about the kinds of communities they were building, or, more narrowly, what those exerting moral influence wanted us to believe about ourselves. Have these three regions been, in turn, the most important seats of the most influential vehicles of propaganda, for textbooks, popular literature, and cinema (and perhaps others)? That question connects us once again with the problems posed in the second of these six major questions, of how actual landscapes became symbolic, which does appear to be a promising entryway into this kind of broad inquiry.

However we might address this general question, it seems a central one for anyone interested in pondering regionalism as an important feature in the course of our national development. A corollary of the question is, of course, why are other regions not represented? What does it say, for example, about Megalopolis—that great urban-suburban-exurban strip along our Atlantic Seaboard that has been a seat of such enormous power and influence in so many realms of American development?

But that question might best be considered within another view of these three landscapes as a *set* of symbols. They constitute a threefold *set of kinds of places:* village-town-suburb. And it is obvious that something important is missing: there is no city, no fully urban, or metropolitan landscape in this set (and sequence) of idealizations. Recall that Suburbia is to be seen not as sub-urban, but as a distinct, separate kind of community.

Of course America has created some world-famous urban landscape symbols. That whole breath-taking panorama of Manhattan is the most obvious example. But while that symbol has long served our national pride, it is a symbol of power, energy, daring, sophistication; but not a symbol of an attractive landscape for American family life.[30] Manhattan in particular and our great urban centers in general have proved attractive to highly selective migrations: to people seeking "success" as defined in terms of money, power, and prestige; or to the poor, domestic and foreign, hoping to find some niche and social support in the vast warren. But for most Americans, the old cliché "a nice place to visit, but I wouldn't want to live there" has expressed their feelings well, and in recent years they are much less interested in or even fearful of an occasional visit.

It is widely claimed and I think deeply true that Americans in general have been and still are strongly antiurban in their emotions. Sam Bass Warner, Jr., notes in his recent book the "long tradition . . . the endless failures of American to build and maintain humane cities" and asserts that "Americans have no urban history. They live in one of the world's most urbanized countries as if it were a wilderness."[31] We are still escapists at heart. When it comes to attractive symbolic landscapes, our propaganda features the "wide open spaces," "Marlboro Country." Our counterculture movements show the still strong attraction of Arcadia, back to the simple country life, the commune in the woods, ways to drop out of the metropolitan maelstrom.[32]

Over many years we have been offered various visions of a better urban future, but the "garden city" seemed to turn into more suburbia; "urban renewal" became all too often a brutal destruction of landscapes and communities in favor of offices and parking lots; technocratic projec-

tions of tomorrow seem to display a sterile environment fit more for electronic robots rather than living, breathing human beings; Soleri's complex
miniaturized city may seem an inspired step toward the Civitate Dei to his
followers but more an anthill in the desert to onlookers.[33] We seem to be an
ever more urbanized people without a sense of direction, without an effective symbol of what the good urban landscape and society might be like.

We may be living therefore in a serious discordance, but we are not
at an impasse. America remains an unusually dynamic and creative society,
and therefore the appropriate question with which to conclude this exploration is:

WHAT IS HAPPENING? IS ANY NEW PATTERN DISCERNIBLE IN THE LANDSCAPES
OF AMERICAN COMMUNITIES? Perhaps the first thing we would notice in a general look at any American city is that people are still moving outward to the
suburbs and beyond. Before assuming that this must mean either that the
critics were wrong or suburbanites remain insensitive to what was being
criticized, we should consider that it may be that there have been significant
changes in the substance of suburbia.[34] There was a period when many a
resident agreed with the social critics: suburbs could indeed be deadly dull
and stultifying if people tried to find a satisfying community life within their
own immediate neighborhoods. The sheer mass and homogeneity of the landscape and society, the distant separation of residence and work, of household
and shops, of home and recreation centers, of sex roles and of age groups,
provided ample basis for the frustration of residents and castigations by
critics, especially in those enormous suburban sprawls of post-World War II.
But by the mid-1960s, when two cars became the minimum family standard
and the engineers had spun a web of superhighways through, around, and
radiating from every city, "community" was no longer a discrete neighborhood, it was a wide scattering of places bound together by the freeways. The
places of sleeping, eating, drinking, relaxing, working, and shopping might
be fragmented among a dozen points separated miles from one another. Thus
we come to the ineluctable observation that the key landscape symbol in late
twentieth century America is not the home but the highway, and community
is not so much a discrete locality as a dispersed social network traced on the
landscape by the moving automobile.

In many ways the automobile rather than the house seems the most
powerful instrument and symbol of our basic values. Through it we express
our individualism, status, freedom, love of mobility and change, as well as
our search for security. It carries us effortlessly to all those amenities and

services made familiar and profoundly democratic by the nationwide uniformity of the McDonalds, Holiday Inns, and a hundred other franchise operations. We move along a linear landscape, intensely developed strips and open interstate routes, made secure and legible by uniform road designs and standardized emblems. And now with citizens band radios, we can communicate directly and anonymously with other individuals, disembodied voices broadcasting at random our craving for contact, as isolated in our steel shells we cruise the monotonous uniformity of the interstate highways.

But a nagging question hangs over this scene: can this kind of atomized dispersal of people living in motorized and electronic connection with their environment and with one another be called "community"? Certainly many would call it a travesty of such. And so the great American longing for something more humane, intimate, stable, and satisfying goes on. The "search" for community, the "quest" of community, the "eclipse" of community persists as a major theme in American thought.[35] *Our Town* was an explicit response by Thornton Wilder to that quest, a conscious idealization to counter the reality he saw of the American as "a nomad in relation to place, disattached in relation to time, lonely in relation to society."[36] Despite its obvious power in fact, despite the great American phenomenon of the mobile home and the motorized home, despite the power of the romantic image of the uncommitted footloose traveler, the Easy Rider drifting from one pad to another, most Americans would not be comfortable with the highway as the appropriate symbolic landscape for a satisfying concept of community.[37]

Not only is there no obvious symbolic landscape of American community today, there is no clear image or even simple common terms for the kind of setting and society most Americans live in. It is not "urban" in the most common sense and symbolic references of that term. The encompassing unit is a metropolitan system of diverse parts, including old densely urbanized areas, suburbs of various ages and character, engulfed towns, roadside strips, shopping plazas at beltway interchanges, a wide variety of discrete residential tracts, former hamlets, towns, farms, and all manner of individual shacks, cottages, mobile homes, houses, and estates scattered over the countryside. The whole complex is bound together by intricate circulation and communication systems and it pulsates with an intense and intimate daily life. It is so vast and variegated and changing that it is difficult to define, but its residents have an intuitive and empirical understanding of its essential character. The highway *is* the fundamental structural element of this landscape, and residents must become travelers skilled in routes, landmarks, and estimates of travel time. In common parlance, going to "the mall" may be replacing going "downtown," but a sense of place with reference to many features may be rather vague and confused, including response to the question "Where do you live?" Insofar as people think in terms of "community," it is likely to involve more than one part of this metropolitan area and the particular combination of parts may differ greatly among residents of the same street. Thus the concept, rarely sharp even in the simplest of settings, becomes ever more blurred socially and geographically, and it is not surprising that we have no ready translations from the realities of such settings into evocative positive symbols. What we have are no more than fragments, most especially the generalized image of the shopping mall, and the upper middle-class suburb with its special style of residences and country club amenities, standardized to cater to executive families transferred every few years from one metropolis to another by national corporations.

It is clear that symbolic landscapes of the character and power we have been considering are not simply designed and marketed to an awaiting public. They arise out of deep cultural processes as a society adapts to new environments, technologies, and opportunities and as it reformulates its basic concepts related to family, community, and the good life. Such changes do not come quickly, and at any particular moment we are likely to be more impressed with continuities than with marked departures from past patterns. Certainly the individual home in the midst of an ample lot, with ready access to a major highway, remains the most prominent element in the landscape of domestic life in America. The relentless outward spread of our cities is powered to a large degree by the emotional bias of an antiurban people caught in an ineluctably urbanizing system. But a new element has rather suddenly emerged within that archetypical domestic landscape: the townhouse condominium. Modern counterparts of the archetypical urban row house, the great majority of these have been built in the suburbs and their acceptance as an alternative to "the free-standing, owner-occupied house of mythologized 'suburbia' involves changes in the symbolism of the house as well as some changes in the functional organization of residential neighborhoods."[38] Such developments reflect important social as well as demographic and economic changes, and while the townhouse by itself may not denote a fundamental shift, it is the kind of landscape clue to broader movements which we may expect ultimately to have significant impact.

It is widely agreed that the past decade or so seems to mark the onset of portentous qualitative changes in American society even though it is hardly possible as yet to define very precisely what these are or to suggest their probable power and trajectory. At best we can perhaps look for specific signs of change and try to assess their meaning within limited contexts. In geography it is axiomatic that social movements do not upwell all across a continent with simultaneous and uniform effect. Rather they arise in a particular place or kind of place, and are subsequently diffused unevenly. We may expect that the national landscape in the future will be shaped, as in the past, by the influence of a pattern which can first be identified as typical of a particular region.

One Possibility

If we search the surface of America today for major regions of change, rather than for simply elements of change such as suburban townhouses, we will have our attention drawn to portions of what journalists have recently dubbed the Sun Belt, a regional concept which encompasses so much fundamental diversity as to be of very limited utility as a framework for careful geographic analysis. Certainly the South has undergone transforming social, demographic, and economic change, including a vigorous expansion of cities and an extensive reorganization of the countryside. But the overall result appears to be best characterized as a belated integration of the region into the mainstream of national life rather than offering any really new direction. Urban and suburban developments and problems seem basically to reflect those common in recent years to all the nation, modified by the deeply-rooted rural tradition of the South which has resulted in a somewhat looser scattering of homes and industries and a more quasi-rural suburban life-style.[39] Houston, at the western corner of the South, is intensely modern, but also not fundamentally new. Here great wealth and vitality have produced the most splendid urban strips, shopping malls, entertainment palaces, office clusters, and residential enclaves, all served by the most efficient freeway and parking facilities. But all of this is essentially a projection of well-established patterns, a display which is authentically Texan in scale, verve and taste, and deeply national in the social values and technology it represents. Houston of the 1970s represents a significantly amplified center in the geography of American corporate power, but it has little national impact as a distinctive and creative cultural landscape.[40] And something of the same can be said for other major sectors of the Belt, such as Florida, Central Arizona, and Southern California. These have been areas of great growth, but in patterns largely reflective of national trends that, in terms of life-style and landscape, had been strongly influenced by Southern California during the period between the World Wars. However if we look just a bit farther, we come to an area which has not only undergone very extensive growth but has been the vortex of a variety of social movements which have so directly challenged national patterns as to become loosely labeled as the "counterculture." The search for new centers of cultural creativity inexorably brings us into a focus on San Francisco, the Bay Region, and, broadly, Northern California.

Since Gold Rush times California has been "the Great Exception," to borrow Carey McWilliams's phrase and argument,[41] a vigorous creative area which has always done many things which differ from the American norm. From the first it had considerable influence upon neighboring regions of the Far West and Pacific Rim, but not until Southern California emerged as the hearth of the new automobile-suburbia culture did it have a strong impact upon the nation as a whole. All during the course of its first century the City by the Golden Gate came to have great symbolic power, but always as a unique place, "everybody's favorite American city," loved because it seemed so different from the general character of American cities, attractive especially to those who sought an alternative to the common patterns of American society and landscape.

However, during World War II and its immediate aftermath San Francisco rather suddenly became much more closely articulated to the nation. As Vance has noted, "the founding of the United Nations in the San Francisco Opera House in 1945 . . . symbolized the integration of California into the heartland of ideas and actions," and we can draw heavily upon his seminal exposition of the decades which followed during which "in a fascinating cultural-geographical isostasy, the Bay Area, California, and the West Coast have risen in their impact on American settlement structure as New York and Megalopolis have sunk."[42] The mere mention of names such as North Beach, Haight-Ashbury, Berkeley, Big Sur, the Black Panthers, the Sierra Club, and, indeed, Governor Jerry Brown suggests something of the range of movements which have reverberated through the nation. The Bay Area has been a fertile seedbed of new art, music, literature, religious expression, psychological exploration, and educational experiment; it has been the principal seat of the ecological and environmental movements and of challenges to national attitudes toward unlimited growth, consumption, and technological proliferation; it has been the most famous setting for the assertion of new patterns of individual and group consciousness and of a great variety of experiments in alternative styles of life. Although many of the specific expressions of these movements have been merely sensational, superficial, ephemeral, or parochial, there is no doubt but what they have deeper levels that do represent a critique of some of the fundamentals of the American way of life, and no doubt that they presage important social change.

Vance suggests that "we are witnessing the birth of a new complex urbanism in which the specialized social districts have begun to replace a synoptic pattern (of land rent) in shaping the morphology of settlement."[43] He sees the Bay Area as being gradually reorganized through a self-sorting of people not by class or income, nor even very firmly by ethnicity or race, but by life-style, resulting in "voluntary districts" (to use Zelinsky's term[44]) formed out of the search for a way of life which may be quite at variance with what have been the cultural norms. It is conceivable that from such developments San Francisco might shed its old anomalous status and serve as the chief basis for a new generalized concept of urban life featuring attractive townhouse living, the vibrancy of social heterogeneity, a greater appreciation of townscape, a deeper sense of history and of place, and a greater emphasis upon the humane rather than the material aspects of life so that the core becomes increasingly more a central social district than a central business district.

It is conceivable, but far from certain. Even in this region the antiurban bias remains strong and Vance describes at some length how the old American search for the ideal has resulted in a strong centrifugal movement, spreading an essentially cosmopolitan population deeply into the woods and mountains of Northern California, radiating northward into Oregon, creating a new Arcadia which is in fact a far-flung Exurbia, the outermost sector of a new metropolitan society. Vance sees this overall complex as "the city-in-the countryside," a new landscape expression of basically old American ideals, and argues that we should be examining the recent history and social geography of California, seeking "to discover not the economically-oriented normative geography, for which Iowa serves well, but rather the cultural dynamics which will foretell the social geography that may well await us all in the near future."[45]

Whether Northern California is the culture hearth of the next in succession of symbolic landscapes we have been discussing remains to be seen. We still know far too little about it, and the whole complex has yet to be defined in terms sufficiently clear and evocative to serve as the means of powerful symbolic expression. What is certain is that new landscapes, actual and symbolic, are being created, and like those we have already experienced they will be at once a mold and a mirror of the society that creates them. If we are interested in interpreting the nature and course of our national life it might be well to give them closer attention.

Notes

1. Cf. Philip L. Wagner, "Cultural Landscapes and Regions: Aspects of Communication," in *Man and Cultural Heritage, Papers in Honor of Fred B. Kniffen, Geoscience and Man*, vol. 5, ed. H. J. Walker and W. G. Haag (Baton Rouge: Louisiana State University Press, 1974), where, under the subtopic "The Morality of Landscape," he states: "I should like to venture the suggestion that all societies ... everywhere and throughout history have regarded some particular sort of environment as uniquely conducive to the good life, and have labored to create it."
2. Christina Tree, *How New England Happened* (Boston: Little, Brown, 1976), p. 135.
3. Cf. John M. Murrin, "Review Essay," *History and Theory* 11, no. 2 (1972): 226–75, ref. p. 234; this characterization is a quotation from Lockridge with specific reference to his study of Dedham, Massachusetts.
4. The classic introduction to this spread is Lois Kimball Mathews, *The Expansion of New England: The Spread of New England Settlement and Institutions to the Mississippi River, 1620–1825* (Boston: Houghton Mifflin, 1909). There is a large literature on the Yankee impact upon the West, but no general survey of it in terms of landscape. John W. Reps, *The Making of Urban America* (Princeton, N. J., Princeton University Press, 1965), has excellent material on New England and Ohio, pp. 115–46, 227–39.
5. It has recently been suggested that there remains such a preponderant emphasis upon New England studies in our historical writing on communities as to constitute "a kind of 'regional imperialism' "; see David J. Russo, *Families and Communities, A New View of American History*, The American Association for State and Local History (Nashville, 1974), p. 255.
6. James T. Lemon, *The Best Poor Man's Country, A Geographical Study of Early Southeastern Pennsylvania* (Baltimore: The Johns Hopkins Press, 1972).
7. The classic statement of this claim is that of Frederick Jackson Turner, "The Significance of the Frontier in American History," *Annual Report of the American Historical Association for the Year 1893*, pp. 190–227. A seminal essay emphasizing the importance of this Main Street culture of Middle America (along with the New England Town and the Southern County) is Conrad M. Arensberg, "American Communities," *American Anthropologist* 57, no. 6 (1955): 1143–60.
8. Sam Bass Warner, Jr., *Streetcar Suburbs, The Process of Growth in Boston 1870–1900 (Cambridge: Harvard University Press and M.I.T. Press, 1962)*.
9. Carey McWilliams, *Southern California Country, An Island on the Land* (New York: Duell, Sloan & Pearce, 1946).

10. Frank Donovan, *Wheels for a Nation* (New York: Thomas Y. Crowell Co., 1965), p. 158.

11. Howard J. Nelson, "The Spread of an Artificial Landscape over Southern California," *Annals, Association of American Geographers, 49, no. 3, part 2 (1959): 80–100;* James J. Fink, *The Car Culture* (Cambridge: M.I.T. Press, 1975), p. 141; Mark S. Foster, "The Model-T, the Hard Sell, and Los Angeles's Urban Growth: The Decentralization of Los Angeles during the 1920's," *Pacific Historical Review* 64 (November 1975): 459–84.

12. James E. Vance, Jr., "California and the Search for the Ideal," *Annals Association of American Geographers* 62, no. 2 (1972): 182–210; Kevin Starr, *Americans and the California Dream 1850–1915* (New York: Oxford University Press, 1973).

13. Daniel J. Elazar, *Cities of the Prairie, The Metropolitan Frontier and American Politics* (New York: Basic Books, 1970), p. 46.

14. Wilbur Zelinsky, "The Pennsylvania Town: An Overdue Geographical Account," *The Geographical Review* 67, no. 2 (1977): 127. He also notes that the New England village and the Middle Western small town are "perhaps the clearest examples of objects that represent the interaction of community personality and physical substance, at least in the form of durable images in the popular and scholarly minds." (Note 6, p. 128).

15. Murrin, "Review Essay;" Russo, *Families and Communities.*

16. Cf. Zelinsky, "The Pennsylvania Town."

17. Just as this book was going to press, Joseph Sutherland Wood's exemplary study "The Origin of the New England Village" (Ph.D. diss., Pennsylvania State University, 1978) has come to hand to answer these very questions. Wood carefully examines the "conventional" views of New England settlement as a process and as a type, and tests these against an impressive array of evidence region by region within New England. He concludes that very few colonial New England settlements were truly clustered as in the conventional depictions, rather the formation of relatively compact villages took place during a vigorous but brief period of commercial prosperity during the early federal period. However, the concept of community was relatively strong from the beginning and the church or meetinghouse was the focal point of each town. It was around these places of periodic assembly that most commercial villages eventually formed. Until reading Wood, my own view had been grounded upon conventional interpretations, although I had assumed that the basic form was varied and far from stable; I had been impressed with the landscape changes implicit in the analysis of Richard L. Bushman, *From Puritan to Yankee, Character and the Social Order in Connecticut, 1690–1765* (Cambridge: Harvard University Press, 1967).

18. Thornton Wilder, *Our Town, A Play in Three Acts* (New York: Harper & Row, 1938), pp. 5–6.

19. Hildegard Binder Johnson has recently asserted that "public interest, including that of geographers, in a discriminating typology of rural towns in the Middle West has been delayed immeasureably by Sinclair Lewis's novel, *Main Street,* which created for Europeans and Americans alike the 'typical' midwestern town;" *Order Upon the Land, the U.S. Rectangular Land Survey and the Upper*

Mississippi Country (New York: Oxford University Press, 1976), p. 182. Lewis's caricature was of course a harsh critique of the idealized small town. Johnson's book is a fine example of the kind of fundamental historical geographical analysis needed in the study of American cultural landscapes. For Suburbia, the morphogenetic emphasis of James Vance and his students at Berkeley would appear to provide a telling geographic penetration of the common image. See James E. Vance, *This Scene of Man, The Role and Structure of the City in the Geography of Western Civilization* (New York: Harper College Press, 1977).

20. Ima Honaker Herron, *The Small Town in American Drama* (Dallas: Southern Methodist University Press, 1969), pp. 410–15.

21. E.g., Charles S. Aiken, "Faulkner's Yoknapatawpha County: Geographical Fact into Fiction," *The Geographical Review* 67, No. 1 (1977): 1–21. Two prominent British examples are H.C. Darby "The Regional Geography of Thomas Hardy's Wessex," *The Geographical Review* 38, no. 3 (1948): 426–43, and Graeme Whittington, "The Regionalism of Lewis Grassic Gibbon," *Scottish Geographical Magazine* 90, no. 2 (1974): 75–84.

22. It would be interesting to have some samplings of the entire array of landscape depictions over a brief period for these give us important clues to the psychological connections between a people and its land, both in terms of types of environment and of specific regions and localities. See Yi-fu Tuan, *Topophilia, A Study of Environmental Perception, Attitudes, and Values* (Englewood Cliffs, N.J.: Prentice-Hall 1974), for a stimulating reconnaissance of many aspects of such relationships.

23. The topic so phrased of course calls attention to the work of Professor W.G. Hoskins and his associates in Britain. Hoskins's *The Making of the English Landscape* (London: Hodder & Stoughton, 1955), is a landmark study which stimulated a whole series of volumes on counties and regions and a series of television films by BBC. In America, J.B. Jackson, the creator and long the publisher of *Landscape* magazine, has been the chief catalyst; see the concluding essay in this book.

24. *New York Times*, 29 January 1976.

25. The chief spokesman for the opponents of the mill, according to the *Times* reporter, was "the urbane managing editor of *Yankee Magazine*," a popular periodical devoted to fostering an appreciation of the history, society, and landscapes of New England, a potent instrument in the symbolization process.

26. Quoted in George Knox, "The Great American Novel: Final Chapter," *American Quarterly* 21, no. 4 (1969): 667–82. A.C. Hilfer, *The Revolt From the Village* (Chapel Hill: University of North Carolina Press, 1969), is a penetrating assessment.

27. Herron, *The Small Town in American Drama* and *The Small Town in Literature* (Durham: Duke University Press, 1959), offer an excellent reconnaissance of the wide range of depictions through the whole course of our history.

28. Richard V. Francaviglia, "Main Street Revisited," *Places* 1, no. 3 (1974): 7–11. Francaviglia notes the important departure from the prototype (Marceline, Missouri) and the most common form of the Main Street town in that the Disneyland version terminates in a square and a plaza, giving a sense of enclosure and intimacy, a sense of place, which is in marked contrast to the linear thoroughfare of the usual grid-plan Midwestern town.

29. As quoted in Scott Donaldson, "City and Country: Marriage Proposals," *American Quarterly* 20, no. 3 (1968): 547–66.

30. Anselm L. Strauss, *Images of the American City (New York: Free Press, 1961)*, and *The American City, A Sourcebook of Urban Imagery*, ed. Strauss (Chicago: Aldine 1968), are rich in examples. See also Tuan, *Topophilia*, Chapter 13.

31. Sam Bass Warner, Jr., *The Urban Wilderness, A History of the American City* (New York: Harper & Row, 1972), pp. 3, 4. Cf. also Arthur Schlesinger, Jr.'s comments on the deep hostility toward proposals for federal assistance to save New York City from bankruptcy, "Main Street's Revenge," *New York Times*, 29 October 1975.

32. Vance, "California and the Search for the Ideal."

33. Such a glib characterization does violence to the vision and energy which the challenge of creating better urban landscapes has brought forth, but it is not an inappropriate summation within the context of this essay.

34. There is now a large literature on changes over the past 30 years in both the substance and the symbol of suburbia; e.g., *The Changing Face of the Suburbs*, ed. Barry Schwartz (Chicago: University of Chicago Press, 1976), especially the concluding essay by the editor, "Images of Suburbia: Some Revisionist Commentary and Conclusions"; *The Suburban Seventies*, ed. Louis H. Masotti, Annals of the American Academy of Political and Social Science, November 1975; Peter O. Muller, *The Outer City, Geographical Consequences of the Urbanization of the Suburbs*, Association of American Geographers Resource Paper No. 75-2, 1976; *Suburban Growth: Geographical Processes at the Edge of the Western City*, ed. James A. Johnson (London: John Wiley & Sons, 1974).

35. The phrases are titles of prominent books on the theme: *The Search for Community in Modern America*, ed. E. Digby Baltzell (New York: Harper & Row, 1968); R. Jackson Wilson, *In Quest of Community: Social Philosophy in the United States, 1860–1920* (New York: John Wiley & Sons, 1968); Maurice R. Stein, *The Eclipse of Community, An Interpretation of American Studies* (Princeton: Princeton University Press, 1960). Cf. The opening sentence of Herron, *The Small Town in American Drama*: "in the course of more than three centuries of developing literary expression in America, two characteristics have appeared repeatedly: a tendency toward introspection, and a longing for community."

36. Quoted from his Harvard lectures in the lengthy obituary in the *New York Times*, 8 December 1975.

37. The literary evidence in support of such a conclusion seems very strong. Books, magazines, and newspapers recurrently echo the theme. An almost random example is the "open letter" by Orville Schell in the *New York Times*, 24 February 1977, under the headline "You Can Move On and On. 'But Somewhere It Must End,' " in which he describes how "the word 'community' is a powerful one for me," how he found it in a small California town ("the concept of the small town is one of the most cherished in American history"), and what it requires for survival: "above all, people will have to decide that they have had it with 'moving on.' "

38. Dennis J. Dingemans, "The Urbanization of Suburbia: The Renaissance of the Row House," *Landscape* 20, no. 1 (1975): 20–31.

39. Which is not to say that the South has not been creative in other ways. It has been a powerful force in literature in part because, as Vann Woodward has noted, its regional experience has been anomalous with America but analogous to that of much of the world; see Vann Woodward, *The Burden of Southern History* (Baton Rouge: Louisiana State University Press, 1960).

40. Ada Louise Huxtable, "Deep in the Heart of Nowhere," *New York Times*, 15 February 1976. For a brief assessment of its historical role in Texas and the nation see D.W. Meinig, *Imperial Texas, An Interpretive Essay in Cultural Geography* (Austin: University of Texas Press, 1969).

41. Carey McWilliams, *California: The Great Exception* (New York: Current Books, 1949). See also James J. Parsons, "The Uniqueness of California," *American Quarterly* 7, no. 1 (1955): 45–55.

42. Vance, "California and the Search for The Ideal," p. 205.

43. Ibid.

44. Wilbur Zelinsky, *The Cultural Geography of the United States* (Englewood Cliffs, N.J.: Prentice-Hall, 1973), pp. 134–39.

45. Vance, "California and the Search for the Ideal," p. 204.

IV

Teachers

Reading the Landscape

An Appreciation
of W.G. Hoskins and J.B. Jackson

D.W. Meinig

Introduction

"A rich and beautiful book is always open before us. We have but to learn to read it." Thus in the Spring of 1951 did J. B. Jackson of Santa Fe, New Mexico, suggest his intentions in the first issue of his new magazine, *Landscape*.[1]

"The English landscape itself, to those who know how to read it aright, is the richest historical record we possess." Thus in 1955 did W. G. Hoskins of Oxford University assert the significance of the theme of his new book, *The Making of the English Landscape*.[2]

Neither of these new publications had much immediate impact, but we can now look back and see that their appearance at mid-century marks the beginnings of a new era in the history of landscape studies in the English-speaking world. Each proved to be the first item in a continuously expanding body of work by a vigorous and penetrating student. Each man has looked at the landscape in new ways, defined a new focus for its study, and has through his own work and his stimulation of others created a new literature. Each has had a major influence in his own country, and an impact upon more than one field, and each has consciously sought to reach out to people well beyond the usual professional circles.

Yet it is no contradiction of these general statements to say that the specific character, the range, and some important implications of the work of Hoskins and of Jackson are not yet widely understood. Furthermore, partly for reasons peculiar to their special interests, they certainly are not equally appreciated on either side of the Atlantic. Although they have in many ways played analagous roles and they speak alike in calling for us to open our eyes and see the landscape in new ways, they are in fact very different men who have sought to open our eyes to different things for different purposes. It is my hope that this brief review and assessment will help us better to appreciate what they have done, to see how they are alike and how they differ, and to consider how we might draw upon the ideas of both in the continuing development of landscape studies.

The Making of the Landscape

William Hoskins could quite properly call *The Making of the English Landscape* "a pioneer study." Although books about the landscape, scenery, and the countryside, about localities and "topography," had long been a staple category of British publishing, none dealt with the detailed history of the ordinary man-made scene. Furthermore, not only was it "a new kind of history" which, if its claims were accepted, would require a considerable shift in focus and methods in the profession of history, bringing many researchers out of the archives and into the field, it was also a kind of history which sought to reach well beyond narrow professional concerns to bring a new kind of pleasure to the more general realm of landscape appreciation. Indeed, it aspired to transform landscape viewing into a humane, historical art:

> One may liken the English landscape ... to a symphony, which it is possible to enjoy as an architectural mass of sound, beautiful or impressive as the case may be, without being able to analyse it in detail or to see the logical development of its structure. The enjoyment may be real, but it is limited in scope and in the last resort vaguely diffused in emotion. But if instead of hearing merely a symphonic mass of sound, we are able to isolate the themes as they

enter, to see how one by one they are intricately woven together and by what magic new harmonies are produced, perceive the manifold subtle variations on a single theme, however disguised it may be, then the total effect is immeasurably enhanced. So it is with landscapes of the historic depth and physical variety that England shows almost everywhere. Only when we know all the themes and harmonies can we begin to appreciate its full beauty and discover in it new subtleties every time we visit it. . . . This book is . . . an attempt to study the development of the English landscape much as though it were a piece of music, or a series of compositions of varying magnitude, in order that we may understand the logic that lies behind the beautiful whole.[3]

That last phrase, "the logic that lies behind the beautiful whole," declares the bold attempt to engage both reason and emotion, to combine explanation and evocation to give a "many-sided pleasure" to the skilled historian and sensitive traveler.

The book is structured as history, a chronological sequence of general eras. After a brief attempt at imagining the pristine scene and "trying to enter into the minds of the first men to break into a virgin landscape," there follows a short review of Celtic and Roman remains, and then a major emphasis on Anglo-Saxon settlement as forming the main basis of the design to be seen on the land in the 1950s. Subsequent chapters deal with the great work of medieval colonization, the impact of the Black Death and recovery from it, the flowering of rural England from Tudor to Georgian times, the transformations under parliamentary enclosure, the industrial revolution and a special chapter on roads, canals, and railways, then a separate chapter on the landscape of towns. The book concludes with a short and caustic commentary on "the landscape today."

Such a sequence was in general concordance with those of economic history, but the perspective and emphasis was not for the evidence and the problems were derived not from archives but from the details of ordinary landscapes which lie all around us. The origins of the book go far back to his schoolboy days when Hoskins first felt intimations of the significance of such mundane features. As he wrote in 1973:

> Fifty years ago I used to spend my holidays in a then remote, and still beautiful, bit of Devon. There was no village, only the squire's house (and the Fursdons had been living there since the time of Henry III and were still there), the church and the smithy and the school, and beyond that a landscape of scattered farmsteads, shining in the sun in their green combes for they were mostly built of cob and then whitewashed. The lanes wandered between high hedgebanks, turning abrupt corners and just wide enough for one cart; the fields varied in size and shape but usually made very irregular patterns.

Even then I felt that everything I was looking at was saying something to me if only I could recognize the language. It was a landscape written in a kind of code.[4]

He went on to describe how he had pored over documents and books on local history but none seemed to contain the key to that code. After taking a degree in economic history at Exeter he taught for a year at Bradford and then joined the faculty of Leicester University College. This new acquaintance with the landscapes of Yorkshire and the Midlands raised more and more questions for which no ready answers were available. Gradually he came to see that "in the end . . . I should have to write my own book and try to produce my own answers."[5]

The book began to take shape from extramural lectures on the English landscape which he initiated at Leicester and from his detailed investigations in the Midlands countryside. Hoskins developed a close friendship with the Principal of the College, F. L. Attenborough, who often joined him in the field. Attenborough was a keen photographer and would drive Hoskins to a locality and photograph scenes while the historian carried out his meticulous detective work on foot.[6] Following the long interruption of the Second World War, Hoskins returned to Leicester but found himself increasingly uncomfortable within that academic structure. Attenborough was anxious to retain him and was open to new ideas on directions for postwar development of the university. The result of their discussions was the creation of a Department of English Local History, the first of its kind in Britain. Hoskins was appointed Reader and took up his duties on January 1, 1948 with complete freedom to pursue his landscape and local history investigations. His achievements soon brought an invitation from Oxford to become Reader in Economic History, a post he accepted in 1951 and held for fourteen years.

Hoskins was a pioneer in landscape history. His work can be seen as part of a more general revival of professional interest in local studies related to the development of an array of new techniques and approaches. Hoskins himself gave a succinct description of this larger context in an essay entitled "The Rediscovery of England,"[8] wherein he traced the emergence of a rich local and topographical literature in the late sixteenth century ("the discovery of England") which culminated in the late nineteenth century, whereupon there was a rather sudden decline and shift of interest to other directions. Beginning in the 1920s, the work of new specialists in such topics as place-names, archaeology, architecture, and Domesday geography began to rescue local history from its narrow focus on politics and the gentry.

The Making of the English Landscape can now be seen as a major landmark along the way in this "rediscovery of England," but it was not recognized as such at the time nor did it stimulate an immediate spurt of related studies other than the four county volumes which had been undertaken as coordinate works under the editorship of Hoskins.[9] Hoskins had made a radical claim for the landscape as "the richest historical record we possess," but his fellow historians did not seem to take much notice. The book was ignored by the *Economic History Review.* The *English History Review* gave it a "short notice." The reviewer recognized Hoskins as "a lineal descendant of the great topographers" and as a man with "an exceptionally seeing eye" and "an ample and developed scholarship," but although he wrote that there was much in the book that would be "of service to the scholar" he thought its appeal would be primarily to "a wider public."[10] To serve both the scholar and the public was Hoskins's avowed intent, but one senses here a degree of condescension toward such a book. To call a man a topographer was to place him well away from the mainstream of historical scholarship. Nor was there much indication that the wider public was awaiting just such help for its enjoyment of travel. Although newspaper and magazine reviews were very favorable ("one of those rare books that can produce a permanent and delightful enlargement of consciousness"—*New Statesman and Nation*), sales dribbled along and the general book and the four county volumes gradually went out of print. In retrospect both the publisher and Hoskins explain this disappointing result in terms of poor book design and lack of aggressive promotion. We might better see it, perhaps, as the uncertain interval between seedtime and visible growth in a trial planting of a new kind of literature.

Meanwhile Hoskins persevered in his intensive local investigations, and in 1965 he was invited to return to Leicester to what was now the Chair of English Local History.[11] He accepted that post but was soon restive with academic bureaucracy and in 1968, at the age of 60, he retired to devote himself full time to research and writing and to foster further work in landscape history. There followed a surge of activities and products which can perhaps best be understood in terms of a coincidence of timing and freedom: that is, of seeds planted in the 1950s beginning at last to bear fruit just when he had freed himself of academic constrictions and competitions. These latter were real and inevitable. Neither landscape history nor local history fit comfortably within the usual pedagogical categories. Any insis-

tence upon the fundamental importance of such microhistory would bring
strong counter arguments from those who worked at different scales and on
broader problems. Any attempt to serve both scholars and a wider public,
and to combine explicitly reason and emotion, would be considered tainted
work by certain kinds of professional colleagues.

Whatever the explanation, the Emeritus Professor became more
heavily and variously involved in landscape studies in the 1970s than he
could probably have envisioned in the 1950s. In 1970, *The Making of the
English Landscape* was republished as an attractive Pelican paperback,
readily available for both public and academic use.[12] In that same year new
books on Dorset and the West Riding were issued as the first in a revival of
the county series by the original publisher, but now completely redesigned
and vigorously promoted. Eight more county volumes appeared in the next
five years.[13] In 1977, *The Making of the English Landscape* was republished
in hardcover in the new series format.[14]

Even more remarkable than this sudden spate of books was an out-
reach into a powerful new medium. In 1972 Peter Jones, a producer at BBC,
came across a copy of the original book, seventeen years after its publication. (Hoskins noted that "authors are accustomed to this sort of time-lag".)
Jones had recently done a film on glacial moulding of the landscape for
Horizon, a science series on BBC-2 Television, and he saw at once the possibilities for a natural sequel. His quick grasp of the book's themes allowed
him to establish a close rapport with Hoskins, and together they undertook
the creation of an introductory film on the making of the English landscape.
Hoskins selected the localities to be featured, roughed out a script, served as
narrator, and appeared in the film as the viewers' personal guide, describing the topics and pointing out the evidence for a wide sample of interesting
problems in landscape history. Jones directed the filming, edited it into the
appropriate length and sequences, selected and adapted the background
music (all by English composers). The finished product was a 51-minute
program first presented in the *Horizon* series at 8:10 P.M., Thursday, July 12,
1973. An estimated one million viewers and several hundred letters in response were sufficient to encourage the making of a series of shorter films
(25 minutes), each focused on the special landscapes of a particular region,
a project completed in early 1978.[15]

Thus we can now describe a remarkable harvest from that one book
of 1955: a paperback edition which has gone through six printings; a new
hardcover edition which became a selection of the Literary Guild and of the
History Book Club and which has sold approximately 27,000 copies in the

William G. Hoskins on location in his beloved Devonshire countryside for BBC-TV filming of "The Making of the English Landscape," September 1972. (W.G. Hoskins)

first year; sixteen county volumes, including thirteen in the revised series (with more in preparation), each issued in printings of about 6,000 with several already reprinted; a special book, *English Landscapes*, an attractive large paperback prepared by Hoskins and published by BBC as an accompaniment for its television presentation which quickly sold out and has been reprinted several times;[16] the original general film which was shown twice on national television and is available on rental to schools and groups; and twelve regional films, each of which has or will be shown on national television and will also be available for further local use. Nor is this quite all. In the Introduction of his 1955 book, Hoskins stated forthrightly his hope to appeal "to those who like to travel intelligently," and as an adjunct of his landscape studies in the East Midlands he authored the volumes on Leicestershire and Rutland in the very popular Shell Guide series. The attractive county books in this series, edited by John Betjeman and John Piper, are composed chiefly of a richly illustrated gazetter listing every village, town, and city, with most of the commentary focused on churches and other buildings of special architectural or historical interest. As we might expect, the two prepared by Hoskins give rather greater attention than most to broader landscapes, village morphology, vernacular architecture, and to topics of special historical interest such as place names and the sites of lost villages.[17]

It takes no more than a reconnaissance of the shelves of any major bookshop in Britain to sense that interest in landscape as history is flourishing in the 1970s. It is an integral part of many of the best in the multitude of studies in local history, it is richly cultivated in historical geography, archeology, the rapidly expanding field of industrial archeology, architecture, and urban history. Nearly every major publisher has some sort of series involving landscape and history, and an especially impressive feature is the fact that such a large proportion of this great volume and variety of books is designed to appeal to a readership far beyond any body of professional specialists. Of course there is much variation in content, mode, and technical level, but it is essential to emphasize that this overall range of variation is a continuum. That is to say, there is no obvious separation between professional and public literatures; it all rests on a rapidly expanding base of sound scholarship.[18]

Of course we cannot credit all this burgeoning literature, even that most directly focused on landscape history, to the influence of one man. Broadly viewed it is all representative of a rich tradition in British scholarship and publishing, and a closer study of it would identify numerous influential writers from a variety of fields. Nevertheless, I think that William G. Hoskins is unrivaled in individual impact. He was a genuine pioneer, the author of the most seminal volume, and as lecturer, tutor, author, editor, radio broadcaster, and combined historical expert and personality on television, he has had the broadest range of influence. Moreover, no one has more consistently projected the reciprocal satisfactions of landscape analysis as a form of history and historical understanding as a form of landscape appreciation. He, more than any other, has blended explanation and evocation in pursuit of "the logic that lies behind the beautiful whole."

But let us now look more closely at this corpus of literature and film. What are the basic characteristics of the Hoskins approach to the analysis and appreciation of landscape?

It is, first and last, a form of *history*. To focus all our attention on *The Making of the English Landscape* and its derivatives is to deal with only half the man. Hoskins is an important English historian whose intensive studies of Devon and the Midlands and most recently of Cornwall and of the early Reformation have dealt with many topics which are not very closely related to the landscape.[19] Thus for Hoskins, landscape analysis is a specialized subfield of history, one whose fundamental importance he has ever insisted upon, but certainly not an activity to be equated with the breadth of history. And thus his landscape work bears the obvious marks of the historian: organized chronologically to deal with a sequence of landscape changes, and much concerned with questions pertaining to origins, sources of influence, characteristics of society in past times, variations in space and time, successions, continuities and discontinuities. He has made clear in various ways that for him landscape appreciation derives primarily from historical understanding, that he is interested in historical problems, not scenery. A corollary is the relentless search for facts, for the hard empirical data for doing history. Thus in 1955 he did not linger over the attempt to envision the pristine landscapes of England; it was a topic logically basic, but about which there was little that was reliably known, and "unless the facts are right, there is no pleasure in this imaginative game."[20]

Secondly, his landscape analysis is always a study of *localities*. The concern for factual detail, the search for evidence visible in fields and hedges, lanes and streets, buildings and clusters of buildings, sets all of his work within a certain scale. For him landscape analysis may not be the whole of local history but it is necessarily a *local* form of history. "The beauty of local history, and of the fieldwork connected with it, is that it involves so much close detail,"[21] and much of that detail must come from walking over the ground time and again, for "no landscape history can be written from documents and other men's books".[22] Certainly for him, stout boots and a good bicycle have been essential tools of research. But there is more involved here than just the personal pleasures of working at this scale: "all the pleasure and the truth lie in the details."[23] To assert that the historian must seek for truth at this scale is to take a stand at one extreme end of a spectrum and to engage in the perpetual disputations over kinds and levels of "truth." By working at this scale, paying attention to every form of human record ("a hedge-bank, a wall, a street, a headstone, a farmhouse, an old man's gossip, wills and tax assessments and the thousands of other written things, a memorial tablet on a church wall") and keeping ever in mind that behind all these records there were people, individual personalities ("we must always be able to hear men and women talking"), the local historian is, ultimately, in Hoskins's words, "trying to restore the fundamental unity of human history."[24] But if only a small proportion of the work of professional historians is set at this intensively local scale, there must be other views which challenge such assertions. We cannot explore here the question of what constitute "the intelligible fields of historical study," to use the phrase of Toynbee, who worked at and argued extensively for the merits of the other extreme end of this scale.[25] One might simply say that the issue surely depends upon the nature of the problems addressed and that local, regional, national, and broader histories ought to be complementary and interdependent, but it would be a glib statement which ignores the complexities of actually "doing history," and ignores the fierce competitions and jealousies endemic in academic guilds. Local history did not have wide respect as a scholarly field when Hoskins began his work, and throughout his career he has felt embattled. Bold statements in its defense are sprinkled widely through his writings, at times in provocatively extreme form, and never more so than in the quotation from William Blake which he placed as the epigraph to his collection of studies, *Provincial England:*

> To Generalize is to be an Idiot.
> To Particularize is the Alone Distinction of Merit.

One senses that the writing of a general book on the English landscape was an uncomfortable task for him. He was reluctant to do the first film on England as a whole, preferring to begin with at least half a dozen films on selected topics in various localities. But the first book and the first film could be accepted as pioneering necessities, and in each case a county series could be anticipated as an amelioration of the discomforts of generalization.

Landscape analysis, then, must be rooted in detail, but there is the inherent problem that there is no end of detail to analyze. One can pay attention to the shapes and patterns of the bricks as well as those of the buildings and of the villages as a whole. Each has much to tell (as architectural historians will readily attest about bricks and buildings). Hoskins is certainly aware of a wide range of detail—he writes about subtle textures as well as gross forms, and about interior plans as well as the exterior shapes of houses—but the main emphasis in his most important landscape studies is upon the *morphology* of a locality. As he wrote, in an essay of 1955:

> The view I see, looking at a wide landscape as an historian, is a view composed in the first place of fields, fields of various shapes and sizes, grouped together in certain ways; a view with open roads or deep-set half-buried lanes;

of churches with or without spires; of grouped villages or perhaps lonely hamlets, or even single farmsteads; of towns of a certain shape and size—in short, everything that has been humanly made, and not sculptured by nature in the first place.[26]

Dismissing the summary phrase as being a declaration of a general realm of study, the view here is basically one of areal composition. Hence the special attention to boundaries, to balks, banks, hedges, lanes, roads, and streets which are palpable clues to land parcels, to cadastral units which tend to persist through generations and even millenia. With this basic organizational pattern ever in mind, one can then examine the sites of major elements, the exact positioning of farmsteads, cottages, great houses, churches, hamlets, and villages. Furthermore, one studies the more general form of these elements and their relationships to one another, always alert for oddities, incongruities, for features which seem to have no rationale in terms of present-day activities. Such things are likely to be vestigial, clues to an earlier landscape pattern reflective of a different set and context of activities. They are means of deciphering the code and it is such detective work and the feeling that one has got it right, that one can visualize that earlier scene and sense the changes which have taken place, that gives great satisfaction to the historian.

For Hoskins there has been the added thrill of sensing a remarkable continuity in so much of the English settlement pattern: "there are places where one treads on three thousand years of history, sometimes four thousand or more."[27] This special sense of time is presumably not uncommon among historians, but it must have a special power for those engaging in such landscape research, for the evidence is so palpable and in many cases so comprehensive for those who know how to read it. No one has more eloquently described the pleasures:

> It is satisfying to sit upon a Saxon boundary bank that commands a view of perhaps three or four miles, no more, near enough for everything to be seen clearly, and to be able to give a name to every farm dotted about the fields, to every wood and lane, to know which of these farms is recorded in Domesday Book, and which came later in date in the great colonisation movement of the thirteenth century; to see on the opposite slopes, with its Georgian stucco shining in the afternoon sun, the house of some impoverished squire whose ancestors settled on that hillside in the time of King John and took their name from it; to know that behind one there lies an ancient estate of a long-vanished abbey where St. Boniface had his earliest schooling, and that in front stretches the demesne farm of Anglo-Saxon and Norman kings; to be

aware if you like that one is part of an immense unbroken stream that has
flowed over this scene for more than a thousand years. . . . [28]

Of course to read the landscape in such a manner requires the help of
other sorts of evidence and perhaps the help of other specialists. *"Everything
is older than we think,"* wrote Hoskins recently, with obvious satisfaction.
And he included himself in the admonition, for he was commenting on how
archeological investigations of the past twenty years have forced him to
revise his own views on landscape continuities. He has warmly welcomed
such testimony from the spade, but he has persisted in limiting his own
conception of landscape analysis to that which can be found on the surface:
"The visible landscape offers us enough stimulus and pleasure without the
uncertainty about what may lie underneath."[29] Such a stance defines his
brand of landscape study as a special form of selective retrospective history:
one begins with the present and is concerned essentially only with those
parts of the past which persist. It also suggests in both tone and concept
how completely he has bound together landscape history and landscape
appreciation. There is pleasure in "sitting upon the Saxon hedgebank in the
afternoon sun, looking over a view as richly orchestrated as a Brahms symphony" only if one understands a good deal about the orchestration, only if
one can "look at every feature with exact knowledge, able to give a name to
it and knowing how it got there, and not just gaze uncomprehendingly at it
as a beautiful but silent view."[30]

Such fusions of history and emotion are characteristic of Hoskins's
work. They are fundamental to its attractiveness and success, but they also
at times reveal some serious limitations. One expects an historian to be
fascinated with the past; it need not follow that he hate the present. Few
have been more consistently outspokenly *antimodern* than Hoskins. Consider the opening paragraph of "The Landscape Today," the final chapter of
his landmark book of 1955, which stands unchanged in the reprints of 1970
and 1977:

> The industrial revolution and the creation of parks around the country houses
> have taken us down to the later years of the nineteenth century. Since that
> time, and especially since the year 1914, every single change in English land-
> scape has either uglified it or destroyed its meaning, or both. Of all the
> changes in the last two generations, only the great reservoirs of water for the
> industrial cities of the North and Midlands have added anything to the scene
> that one can contemplate without pain. It is a distasteful subject, but it must
> be faced for a few moments.[31]

But he is so overwhelmed by what he sees that he can face the subject for only three short paragraphs, and he uses these to denounce this "England of the Nissen hut . . . England of the arterial by-pass . . . murderous with lorries. England of the bombing range . . . Barbaric England of the scientists, the militia men and the politicians," and then invites his readers to "turn away and contemplate the past—before all is lost to the vandals." From that point the book concludes with a beautiful coda, a succinct elaboration on his great themes, as in eight rich paragraphs he details the history which can be seen in the bit of "gentle unravished English landscape" out the window of his writing room in rural Oxfordshire.

It is easy to sympathize with his view: to sense the heartache over the war-battered landscapes of an England ravaged as much by the martial tramplings of her own armies and allies as by the bombs of the enemy; to be moved by his eloquent disgust that peace brought, droning day after day, "the obscene shape of the atom-bomber, laying a trail like a filthy slug upon Constable's and Gainsborough's sky."[32] One can readily nod in agreement over the "crimes" committed by the electricity board with their "hideous poles and wires,"[33] over his preference for an old village inn even of no great quality to "the harsh chromium plate, television, and tubular lighting of the twentieth-century pub."[34] One knows what he means by the "incessant noise, speed, and all the other acids of modernity" and one delights in his reference to the rural backroads of Rutland where "one can go slow almost anywhere without instantly being regarded as an anti-social pest, hooted at from behind and then glowered at as the Other Man rushes by with the Death-Wish written all over his flaming face."[35] One appreciates his contempt for urban theorists devoid of any aesthetic or historical sense who seem to dominate the world of planning, and his feeling that it is the city of Leicester's good fortune to have no real center, "least of all—praise God—a civil centre in councillor's concrete."[36] One undertands and one sympathizes—but to follow him all the way is to abandon landscape analysis and simply indulge one's emotions. Hoskins has declared the twentieth century an abomination to be denounced or ignored. I believe I am safe in saying that in neither his general book nor his detailed study of the Leicestershire landscape is there a single mention of a twentieth century building. As works of scholarship they therefore stand as incomplete; by militantly, idiosyncratically refusing to describe and analyze one of the greatest eras of change, Hoskins has arbitrarily truncated the story of the making of the English landscape.[37]

In part this may be the bias of the historian showing again. So much of the modern is discordant with the past, disrupting those long continuities which so fascinate him—but that was true of the Reformation, the enclosure movement, the industrial revolution as well. Or perhaps it is simply the not uncommon historian indifference to recent events, a feeling that there is no really exciting detective work to be done. After all, Hoskins has witnessed most of this century, read about it as journalism rather than history, and he may feel that he knows and that most of his readers know quite enough about the nature and timing of its landscape changes. And even where one does not really understand all the processes involved, cracking the code would lead one into the boring anonymous thickets of bureaus and councils run by "black-hatted officers of This and That."[38] But we cannot excuse him so easily. We cannot avoid the conclusion that there is a deep sentimental bias laced through all his landscape writing, and while his incisive expressions of it enliven his books and vivify him in our minds as a strong and engaging personality, they also limit our confidence in his judgment and, at worst, depreciate him as an historian.

To insist that "every single change in English landscape" in the twentieth century has "uglified" it is more than an aesthetic judgment. The real issue is that "Demos and Science are the joint Emperors" and these cannot, by their very nature, create anything admirable. To lament the decay of the great country houses and simply ignore new housing in any form, to lament the bulldozing of hedges and ignore the realities of agriculture as a way of life in modern times as well as in the past, to curse the motorcar and all of its thoroughfares[39]—such things, it seems to me, impair both landscape history and landscape appreciation. It is a limitation, for it blots out an era from serious consideration. It encourages us to avert our eyes from looking directly, steadily and inquiringly at an increasingly large proportion of the English landscape, and it raises deep questions about relationships between visual pleasures and historical excitements on the one hand and the realities of social life on the other. Hoskins is an immensely knowledgeable local historian and he is well aware of social conditions in times past, but when he declares "how infinitely more pleasant a place England was for the majority of her people" in 1500 as compared with today,[40] we pause and ponder. The case may be arguable, but his antimodern bias has been so persistently expressed that we must wonder how deeply he has considered the difference between looking at a landscape and living in it; and whether he has really been able "to hear men and women talking" in the times of Henry Tudor and of Elizabeth II.

And yet, despite this indictment, his personal judgments are fundamental to the power and the importance of his work. When we set aside his fulminations against the modern world and let him take us to some mundane scene he has tramped over and deciphered with his eyes and his boots as well as his archival research and listen to him describe and define and share the joys of doing it, we know we are with a master. In all landscape literature no one has equalled his *evocation* of the 'many-sided pleasures" of his kind of search "to understand" and to "savour to the full" the English landscape.

"Poets make the best topographers," he stated in the opening sentence of *The Making of the English Landscape*. He was referring to Wordsworth, but it would be an apt epigraph for his own work, for he can write with poetic efficiency about scenes and feelings. Throughout his career he has drawn upon the power of plain English, and forthrightly expressed his "temperamental allergy" to the "unpalatable jargon of learned trades." He has been remarkably unafraid of being thought sentimental or unscientific by his colleagues, and has shown his contempt for the pretentiousness of so much in academic life. ("I once wrote a book with the simple title of *The Making of the English Landscape*, but I ought to have called it *The Morphogenesis of the Cultural Environment* to make the fullest impact."[41])

We have already heard him define the pleasures to be derived from his kind of historical reading of the landscape, as in that view from the Saxon hedgebank. Let us now listen to him speak of the simple satisfactions to be derived from the perambulations of the sensitive landscape traveler: as, for example, when he concludes his Leicestershire volume with a few scenes remembered with special affection:

> The farmhouse in the fields: the white-sashed Georgian windows shining in the afternoon sun, the post-and-rail fence, the coffee-coloured ploughland, the hawthorn trees in bud, or best of all in full flower, the distant ash-trees feathery in the light wind.

Or,

> One may recall the old bridle-paths that form a green network all over the countryside, shaded by ash and hawthorn, hedged in by blackberries: above all, perhaps plodding along them in a thin autumn rain from one remote village to the next, seeing the twiggy spinneys across the fields, and the sharp blackthorn needles of the church spires silhoutted against the blank November sky; and on and on to the main road as darkness falls and the lighted galleon of the Midland Red bus comes to take one back again to the dim smoky hollow where the city of Leicester lies in its Saturday evening peace.[42]

And there are less pleasant but no less memorable views, as of a late nineteenth century industrial township:

> The sight of South Wigston on a wet and foggy Sunday afternoon in Novem-
> ber is an experience one is glad to have had. It reaches the rock-bottom of
> English provincial life; and there is something profoundly moving about it.[43]

Or, finally, listen to his prescription for readers of his Shell Guide not on
what to see, but simply how to explore and enjoy "the wide, water-coloured
landscapes" of Rutland. The broad grassy verges of its country roads
(created by the enclosure commissioners) invite a roadside picnic

> so one does as in France: buy bread, cheese, fruit, and a bottle of wine in town
> before setting out for the day (no hotel "packed lunches," which are usually
> grisly and not worth half the money charged for them) . . . and a bottle should
> last a husband and wife for two days unless one intends to fall asleep after
> lunch. Red wines are better than white for this form of travel as white wines
> tend to become tepid in a car and therefore rather horrid. . . .
>
>   So one is set for a good morning among the churches: not more than
> three in a morning is sufficient for enjoyment, and perhaps another one or two
> after tea: and the afternoon perhaps spent lying in a field in the Midland sun,
> quietly unwinding. I have found Tixover churchyard a pleasant place for an
> afternoon doze. It is very peaceful: the only sound the sighing of a gentle wind
> along the grasses of the Welland Valley. Then in the evening, the return in the
> last sunshine of the day to Oakham, or Uppingham, or Stamford . . . and, one
> hopes, a good well-cooked English dinner. A week of this treatment in Rutland
> sends one back ready to fight politicians and jacks-in-office with one hand tied
> behind one's back.[44]

The Meaning of The Landscape

The title page of the inaugural issue of *Landscape* gave some indication of
its founder's intentions. The subtitle, "Human Geography of the Southwest," defined it as a regional magazine, and an invitation to authors further stated:

> *Landscape* is interested in original articles of not more than 4,000 words deal-
> ing with aspects of the human geography of the Southwest, particularly those
> suited to illustration by aerial photographs. Articles should be designed to
> appeal to the intelligent layman rather than the specialist.

An opening editorial, three short articles, a section on "Books," another on
"Tourism," and a concluding statement on "The Vocation of Human Geography" completed the announcement of offerings.

The Southwest was of course famous for its colorful scenery and picturesque peoples, a "land of enchantment" long extolled in travel literature. But there were clues here that Jackson's interest in landscape was rather different from that being featured in, for example, *Holiday*, another new magazine of the time. *Aerial* photographs? Aerial photographs tend to suppress scenery and reveal design; they may come to be seen as having a special beauty but they are, first of all, tools of analysis which the general reader would hardly know how to use. *Human* geography? The term appears three times on the title page, but the editor was evidently aware that few "intelligent laymen" in America would have any clear sense of what it might mean, and so he placed at the back a long excerpt from Maurice le Lannou's *Le Geographie Humaine* which described it (in Jackson's translation) as a "straightforward study" of the "concrete" elements of settlements that displayed how "man-the-inhabitant organizes the world."[45] *Organizes the world?* Whatever that might mean, it suggested the need for analysis and study; that to "read the landscape" as Jackson intended would require a special literacy.

Jackson offered a substantial illustration of just such literacy in his own essay entitled 'Chihuahua: As We Might Have Been.'[46] The theme was the cultural significance of the international boundary between Mexico and New Mexico. A hundred years ago "an abstraction, a Euclidean line [was] drawn across the desert" and "now there are two nations, two landscapes, two ways of looking at the world and of living in it." The contrast was concrete, vivid in the landscape; but the cause was "psychological," imbedded in attitudes toward country and city life and social status: "south of the Rio Grande the world of man is thought of as created in the likeness of a social theory and not, as with us, in the likeness of an economic force." It was a provocative essay which linked landscape to philosophy and culture and did so in a clean clear prose devoid of academic jargon. If this was geography, it was a far remove from what was usually found under that name in the classrooms of America. If this was travel literature, it was a far remove from what was usually served the American tourist.

A reader with a professional interest in such matters who happened upon the first issue of this handsomely printed little magazine must surely have been surprised and intrigued. How was it that such a sophisticated view, explicitly associated with French ideas, should appear in an unheralded periodical issued from a post office box address in a small Southwestern city, with no hint of any kind of institutional or professional affiliation? Who *was* J. B. Jackson? And suppose the reader pursued his curiosity, made inquiries in New Mexico, and was told that previous to his new role as publisher-editor-author Mr. Jackson had been a rancher near Clines Corners. *Rancher?*

Now in fact most of the first readers of *Landscape* knew who Jackson was, because, starting out utterly on his own, he simply mailed copies to his friends and acquaintances and to a few persons he thought might be interested; *Landscape* was launched with twenty subscriptions. And if they knew him at all, those readers knew that John Brinckerhoff Jackson was no ordinary rancher. Born in Dinard of American parents, he had been schooled in America, Switzerland, and France; had graduated with a B.A. in history and literature from Harvard (1932); and had spent time in a variety of activities: working on a newspaper in New Bedford, on his uncle's ranch in New Mexico, studying architecture for a year at M.I.T., making annual trips to Europe. After several years service as a combat intelligence officer in World War II, Jackson returned to New Mexico and leased a ranch fifty miles south of Santa Fe. But his intended career was soon abruptly halted when he was thrown and dragged by a horse. A year in the hospital gave him plenty of time to think about an alternative.

The war had forced Jackson into an intensive study of geography at a very local scale. He had made use of maps, photographs, aerial views, guide books, interviews, and anything else that could provide detailed knowledge of what lay in the path of the advancing Allied armies. After the war while stationed near Aachen he had written several little guides to localities for the use of American soldiers, and he continued to steep himself in the rich European topographic and geographic literature. Therein he could see how many of his interests in history, geography, architecture, and ideas seemed to converge and find expression in a kind of descriptive and interpretive literature which had no general American counterpart. As he lay in the hospital he began to consider that he might do something about that. He recalls that the specific idea of a magazine was prompted by the appearance of a new French publication, *Revue de geographie humaine et d'ethnologie*, edited by Pierre Deffontaines.[47] The decision to focus on the Southwest was a natural expression of his love and concern for his adopted region.

Jackson tried to express what it was he wanted to convey about the Southwest in his opening editorial, under a title borrowed from Robert Frost, "The Need of Being Versed in Country Things." The need was to create a far more effective appreciation of the countryside and its problems on the part of an increasingly urbanized society. There were powerful national trends underlying what he saw as an unhealthy and growing dichotomy, but he felt that some specific blame must rest upon broadcasters and editors who routinely gave precedence to "the gossip of Broadway and Hollywood" over important developments in their own regions. There was an urgent need for the public to have a better understanding of how their local communities and countrysides were being changed, and that understanding must be grounded upon some knowledge of how these settlements were created and how they have endured. The Southwest offered a rich variety for study, and air travel had opened up a fresh and attractive approach.

> It is from the air that the true relationship between the natural and the human landscape is first clearly revealed. . . . No one who has experienced this spectacle . . . can have failed to be fascinated by it, nor wonder at the variety of men's ways of coming to terms with nature. Why are some stretches of land thickly settled with villages almost within sight of one another, while others are occupied by great rectangular fields and a few lonely houses? What brought into being so close a network of roads and trails and bridges? Some farmhouses are small and primitive, others have a dozen satellite buildings; and some are so close to their neighbors as to form villages and small towns. And how to account for the many types among the towns themselves? There are the compact Indian communities, perched on rocks overlooking the fields, the sprawling tree-grown checkerboard of the Anglo-American towns, the Spanish villages strung along a road or a stream; the huddle of filling stations and tourist courts at highway intersections in the desert.
>   And to such questions an equally varied set of answers occurs . . . [but] the asking of such questions is more important than the finding of an answer. It means that, like the air traveler, we have acquired a new and valuable perspective on the world of men, and with it eventually comes the realization there is really no such thing as a dull landscape or farm or town. . . . Wherever we go, whatever the nature of our work, we adorn the face of the earth with a living design which changes and is eventually replaced by that of a future generation. How can one tire of looking at this variety, or of marveling at the forces within man and nature that brought it about?[48]

And so *Landscape* was begun "with the hope of arousing a kind of speculative interest in the human geography of the Southwest."[49] And for the next seventeen years J. B. Jackson, through this instrument of his own making, did look so searchingly at the landscapes of man and did speak so clearly and creatively to readers in half a dozen fields that he emerged from obscurity to become an influential and deeply respected figure. Through all those years he was the sole financial backer, publisher, and editor, as well as author of a considerable portion of the content, of *Landscape*. His ideas were on display in many forms: in signed articles and editorials; in book reviews—occasional pieces under a pseudonym ("Ajax") and brief unsigned notes and comments; in the many excerpts which he selected and translated from European literature; in the prepublication excerpts he solicited from the authors of forthcoming books; and in his editorial selection of authors, articles, themes, and the composition of each issue. Thus fifty-one issues (three times a year for seventeen years) of *Landscape* constitute the great *corpus* for the study of Jackson's ideas and the associations he developed with other writers and fields. It is in fact a body of materials too extensive and diverse to be assessed adequately within the space of this essay. It is possible, however, to describe something of the sequence of topics to which he gave major attention, to trace the development of his ideas, and to give a glimpse of his style. From time to time as publisher and editor he paused to review and reassess for his readers the scope and direction of his magazine, thereby creating a set of phases which serve as a convenient scaffold for the making of a survey. However, what follows is not rigidly organized in historical sequence, and perhaps it should also be emphasized that despite the unusually close relationship between the man and the magazine, it is not an assessment of *Landscape* but of J. B. Jackson as revealed in *Landscape*.

The three issues following the first carried numerous articles and reviews on the Southwest, with much emphasis upon vernacular architecture and extensive use of aerial photographs. Jackson offered further models of his kind of human geography in two essays, one a beautifully wrought interpretation of the cultural significance and deep historical origins of the ordinary American front yard, the other a celebration of the virtues and genuineness of the ordinary county seat in the farming and ranching country of the West. [50] In response to a letter from a subscriber who pleaded that *Landscape* not become another sentimental "back-to-the-land" promotion but tell the often grim truth about the problems of country living, Jackson reassured the reader and laid much of the blame for our national ignorance about such matters on "a theory-ridden educational system" which does nothing "to encourage an intelligent and productive interest on the part of the children in local history and geography." [51]

In 1952, after four issues, Jackson announced a shift in emphasis. The magazine had provoked a very favorable response, but there had been insufficient material to sustain the focus on the Southwest. Hence the scope would be broadened and the subtitle changed simply to *Magazine of Human Geography*. Here we see the magazine begin to take on a life of its own, shaped by its contributors and readers as well as its publisher-editor.

Because human geography was in its "formative stage" and remained "comparatively unknown and little understood," Jackson embedded the announcement of this change in a rather lengthy review of the nature and current status of the field. He saw *Landscape* as a complement to the work of professional geographers:

> Our intention is to offer articles, both from American and foreign sources, on those aspects of the human habitat which are most common, most easily perceived, but at present least understood: the dwelling, the town, the road, the rural human landscape.

In that explication he declared a basic principle: *the primary study of the human geographer must be the dwelling."* The house is the microcosm, the prime example of Man the Inhabitant's effort to organize his environment, to create a landscape which will satisfy not only his biological but also his social and spiritual needs.[52] In the next issue he offered an example of what he meant in an essay describing three hypothetical American houses of three different centuries, in the lineage of the same family, illustrating great differences in social values and ways of life.[53] *Landscape* continued to give major attention to rural and Southwestern topics,[54] but a much broadened scope was evident in the articles, excerpts, and reviews relating to modern architecture, planning, highways, and urban developments, with examples and reports drawn from Europe, the Soviet Union, China, Brazil, and Africa. And one begins to suspect a shift rather greater than announced when, in a preface to a long excerpt of Max Sorre's "The structure of cities," Jackson began by stating that "whatever its point of departure, every discussion in the field of human geography sooner or later comes back to the city as the supreme example of man's modification of his environment."[55]

In 1956, after five years of publication, Jackson made another public assessment of the direction of his magazine. He expressed general satisfaction with the broadening of scope, especially the inclusion of much material from foreign literature. He reiterated his concern over the pervasive failure of leaders and the public alike to appreciate the problems of rural and small town America caught inexorably by rapid technological and social change, and he went on to assert his belief that landscapes could and should be evaluated in terms of their qualities for human living, and declared his "very substantial hope that the ills at present besetting the human landscape can eventually be cured" through appropriate planning. However, he made clear that the real grounds for that hope lay not in the field of planning as currently practiced, but in a far more decentralized mode which would be sensitized to the interests of local inhabitants. Thus he saw a role for his magazine in the "struggle for effective landscape planning." *Land-*

scape would try to display something of "the complexity and fascination of the problem" and of the necessity for trying "to experience the landscape in terms of its inhabitants."[56] Although in this "statement of policy" he continued to couch his main themes rather directly in the context of the Southwest, the magazine more and more reflected his wide reading and travels and ecumenical thinking. Synopses of articles currently appearing in foreign language periodicals were regularized and expanded in a special section; occasionally an entire issue was devoted to a foreign area (the Mediterranean, Turkey and the Balkans). Urban problems, architecture, and planning in many different guises became much more prominent.

It was characteristic of Jackson that, having stressed the need for a different approach to planning, he would try to illustrate through his own essays what he had in mind. His readers had already been given a good many clues, most directly expressed in his sharp critique of the doctrinaire views permeating some of the major books of the day. In his review of Garrett Eckbo's *Landscape for Living*, he noted "how abstract the basis of much modern planning has become." He chided Mr. Eckbo for being "more at home in the principles of planning than in the landscape itself," and he declared those very principles to be fundamentally wrong because they assumed that through technology and a reformed social order the landscape could be rationally designed for the Common Man. Jackson warned that such planners appear

> to think in terms of people instead of persons, and the communities they design are often as intolerant of the individual's undisciplined aspirations as of the undisciplined forces of nature. The greatest shortcoming of Mr. Eckbo's philosophy is his contemptuous dismissal of the imponderable qualities which develop not only in every man but in every society.[57]

Similarly, he saw E. A. Gutkind's *Our World From the Air* as a proclamation that "large-scale, long-term planning is the key to our salvation," and he was appalled at the pretentiousness and alarmed at the ignorance displayed, at the "tendency to reduce the complex ethnic, geographical and historical diversities of the human landscape to a few ready made formulae." The intention may have been benign but the implications were grave. The approach was based on "a kind of one worldism" which was in fact a "parochialism of the blindest sort; which believes and publicly states that the differences separating one race and one nation from others are of a trivial and easily explained nature." The aerial view was much to be recommended, but

What we need is not an aerial perspective of the globe but an aerial perspective of our own backyard. It is no use telling us that the world is our home and that we should love it, until we have learned to love our own corner of it, until we have learned what that corner possesses in the way of beauty and potentiality for happiness. Otherwise we can only look on the foreign landscape as Dr. Gutkind does: not as an inhabitant, but as a kind of global social worker, bereft of all sense of wonder, but very much aware of what is wrong and how to right it.[58]

And so Jackson now turned his attention to focus directly upon a part of our "backyard" which architects and planners either ignored or abhorred: the roadside strips which were growing at such a phenomenal rate in America. He readily acknowledged that much of it was a blight which must be brought under control, but he cautioned that before we seek legislation to suppress, select, or impose design restrictions upon such facilities, we had better try to look at them from the standpoint of their owners and users and to reflect upon what it all represents: "How are we to tame this force unless we understand it and even develop a kind of love for it?"

A kind of *love* for all that garish clutter and confusion? Yes, once we learn to see it as part of the exuberant vitality of an America responding creatively to the realities of a new era. We must see the highway not through the eyes of the traffic engineer or economic planner, but through those of ordinary citizens who spend more and more of their leisure time cruising in their cars; and we must see the garish architecture, shrill signs, and insistent lights not as a "longitudinal slum" but as a "kind of folk art" which is "creating and at the same time reflecting a new public taste." Such a view need not inhibit our powers of discrimination, it only insists that we should make our judgments with an understanding of the complex forces which make the highway strip a powerful expression of "a whole aspect of American life." Aesthetic improvement was much to be desired, but it must be made with an appreciation of the legitimate interests and preferences of the proprietors and their customers rather than prescribed by the professionals as the appropriate Landscape for the People.[59]

Next he looked at another part of our "backyard," a sector of our cities (and especially our smaller cities) which was so unrecognized in professional literature that he had to coin a name: "the stranger's path," that line of facilities and characteristic landscapes which leads from railroad and bus stations, those points of arrival from which the stranger must make his way toward the main centers of urban life. It is another of those penetrating essays drawn from long experiences and acute observation, from the experience of Jackson the solitary traveler making his way in scores of cities in the Old World and the New, observing that these often squalid pathways are one of the fundamentals of urban life—and one ignored in professional treatises on urban geography and urban planning.[60]

In the next issue he probed directly into the *experiencing* of nature. He noted the changes in public recreational styles, from the family Sunday country walk from the end of the streetcar line, to the weekend motor trip, to an array of individual sporting activities: bicycling, skiing, sailing, gliding, motorcycling, hot-rodding. Jackson saw the emergence of this last category as an important new era in man's relationship with his environment. Instead of the gentler traits of the casual stroll, the pause for close inspection or for contemplation of the distant view, these activities all involve such rapid movement through space as to give an utterly different kind of experience of nature:

> The new landscape, seen at a rapid, sometimes even a terrifying pace, is composed of rushing air, shifting lights, clouds, waves, a constantly moving, changing horizon, a constantly changing surface beneath the ski, the wheels, the rudder, the wing. The view is no longer static . . . the traditional way of seeing and experiencing the world is abandoned; in its stead we become active participants, the shifting focus of a moving abstract world; our nerves and muscles are all of them brought into play. To the perceptive individual there can be an almost mystical quality to the experience; his identity seems for the moment to be transmuted.[61]

And here again is the voice of experience, of long and varied experience: of the lad who flew in airplanes in France in the mid 1920s, of the college graduate who got his first motorcycle in Vienna in 1933 and set out to explore central Europe, of the middle-aged college lecturer who routinely crossed the continent on his motorcycle in the 1970s. It was the kind of experience which put him in tune with attitudes toward environments which were rapidly gaining popular strength in the latter twentieth century and thus, he insisted, must be understood by those who would plan the landscapes of the future. The Average Citizen can no longer be served Nature through "more pretty parks or carefully preserved rural landscapes or classical perspectives . . . he has to be on the move one way or another, and he has to be made to feel that he is part of the world, not merely a spectator."[62]

There followed a sequence of articles on urban environments which similarly emphasized the need for close, direct experience, and for breaking out of doctrinaire limitations:

> If I were the Ford Foundation I would give lavish fellowships to students of city planning on the following conditions: that for a year they would look at no picturebooks of Brave New Sweden, attend no lectures entitled "Planning for a More Abundant Democracy" (or "Housing at the Crossroads"), cease all speculating about the City of the Future, and that they spend the time instead deep in the heart of some chaotic, unredeemed, ancient city. Preferably Istanbul . . . squalid, smelly, disorderly, exciting and magnificent . . . what marvelous color and variety, what a superabundance of life! . . . For all its sordidness, Istanbul is a city where urban life has created it own forms, and not the other way around.[63]

> . . . the divergence in points of view between the social minded and what might be called the symbiotic minded is a fundamental one. Properly developed, it is the divergence between an established orthodoxy which has ceased to question its basic tenets, and a radically new approach; between those who think of man exclusively as a social economic being and those who think of him in broader terms.
>
> The symbiotic approach to the problem of the city has an extremely simple and familiar premise: It is that man, in addition to his spiritual identity, is part of nature. . . . [it] cannot think of [the city] simply as a kind of abstract cat's cradle of social relationships.[64]

> As a man-made environment every city has three functions to fulfill: it must be a just and efficient social institution; it must be a biologically wholesome habitat; and it must be a continually satisfying esthetic-sensory experience. Up to the present we have given all thought to the first of these. There are signs that the second will receive its due attention before long . . . But the third will be realized only when we ourselves are enlightened: when we learn once again to see nature in its entirety; not just as a remote object to be worshipped or ignored as its suits us, but as part of ourselves.[65]

In all of these essays Jackson was offering unorthodox commentary on American landscapes informed by his own intimate experiences and cosmopolitan breadth. He was now giving major attention to American cities and highways, but, true to his own principle, he recurrently turned to his own immediate countryside for inspiration and example.[66] And so in the ninth volume of *Landscape* he looked once more at the Spanish-American house of New Mexico. Under the admonitory title "First Comes the House," he argued for the closest study of the vernacular architecture of this provincial culture in order to learn

> What a maze of social, technical, geographical, historical, biographical traces even the simplest house can be; to learn that other societies have in their time reached different but quite viable solutions to some of the problems confronting [the architect],"

to learn how to appreciate

> The link between domestic architecture and the culture which produces it; the link between the single house and the community; and . . . how eventually to evaluate architecture of a greater and more enduring kind.[67]

In a "Tenth Anniversary Issue" in the Fall of 1960 Jackson once more reviewed the course he had followed and where he had hoped to go. He readily admitted that ten years ago he had "never supposed that 'The Need of Being Versed in Country Things' would demand so exhaustive a search," and that he had had to abandon many of his notions about such matters. But even if learning to read the landscape was "more of an undertaking than we had bargained for," he held firmly to his basic principles of how it should be done. Even in this urban age the study of rural landscapes should be required in the training for the design professions:

> Communities and landscapes have always been organized into patterns, but often by anonymous forces or traditions. . . . The architect or planner who becomes aware of this wide and ancient field of anonymous folk design will learn to see purpose where previously he had seen only disorder; and he will perhaps also see that his own designs are at least in part the expression of his own inheritance. The rural landscape and the rural dwellings are not to be studied as models for imitation; but they reveal how forms and patterns come into being, and the process is by no means always rational.[68]

Furthermore, despite the difficulties of seeing it so, the city represents the same kind of patterning, the same basic bond between man and his natural surroundings. *Landscape* had consistently argued against the errors of seeing the city as inherently "unnatural" or of trying to soften its character through the introduction of "greenery and synthetic rural graces" ("gestures of sentimental regret"). Instead, *Landscape* sought a redefinition of the prevailing views of "nature" so that instead of seeing a dichotomy between city and country, between the "natural" and the artificial," we would see that

Nature is actually omnipresent in the city: in the city's *climate, topography and vegetation,* that we are in fact surrounded by an impalpable or *invisible landscape* of spaces and color and light and sound and movement and temperature, in the city no less than in the country. What is more, there is a constant action and reaction between ourselves and this environment. . . . We are beginning to learn that the world surrounding us affects every aspect of our being, that far from being spectators of the world we are participants in it.[69]

He stated that we must work toward an effective formulation of this broader *"definition of nature,"* and that in turn will

bring with it not only a new kind of landscape but a new concept of the landscape, a new kind of architecture and a new kind of urban design. Both in the country and the city it will, we believe, produce what we so badly need: something that can be described as esthetic planning, planning for the delight and stimulation of the senses as well as of the spirit.[70]

Thus it was appropriate that this anniversary issue should feature "The Language of Space" by Edward T. Hall, an essay on "William Carlos Williams' *Paterson*" by Joseph Slate, and the editor's review of Kevin Lynch's *The Image of The City,* as well as a set of articles on the Southwest.

Among these last was an essay by Jackson, "The Four Corners Country." This high lovely country marked by spectacular natural colors and sculptures was an important part of his adopted region which he had known well over many years. In view of the uranium and gas and oil booms of the 1950s he posed the question,

Can anything so indefinable yet so precious as the atmosphere of a place survive the sudden impact of thousands of newcomers, millions of dollars invested in construction of every kind, and a transformed economy?[71]

He proceeded to describe something of the great swirl of activities and of their imprints on the landscape. By far the most distinctive feature was the presence of great numbers of "mobile homes" and the rather ramshackle communities formed of them. He recognized that there was much that was ugly about such camps and that there was some basis for the widespread resentment against trailers and their inhabitants, but, typically, he tried to see them in the broader context of American society, to see them as mid-twentieth century versions of a type: the boom landscape, reproducing a kind of exaggerated, "grotesque versimilitude" of its parent society. Trailers were not only a fact of life which must be faced, they were "very precise expressions of important trends in American life": mobility; the preference for informal living in a small community; the successive reduction in the function of the American home to little more than shelter. Given the sudden need for housing in boom areas, trailers were "a sensible and efficient response"; as shelter they were "triumphs of light, durable construction, ingenious planning, and compact convenience."[72]

The trailer camp of the mining and construction boom is of course a temporary settlement, and many of these were already dwindling. As he looked at the really permanent effects of all that intensive exploitation, he concluded that the damage to the landscape was really very slight, and that the precious "atmosphere of the place" did survive. Indeed, in retrospect, "The House Trailer People" seemed to have belonged here no less than The Navaho and a sequence of others.

> They are wanderers in a landscape always inhabited by wanderers. They never settled down. The way they came out of nowhere, stayed awhile and then moved on without leaving more than a few half-hidden traces behind, makes them forever part of this lonely and beautiful country.[73]

From such a stance one could penetrate the clichés, look the twentieth century American landscape squarely in the face, and often find beauty where others saw only despoliation. The most interesting of the trailer courts, he wrote, were the remote satellite camps

> a half hour out on a dirt road in the range or desert, set in the midst of a gigantic landscape of rock and sunbleached grass. You drive through the emptiness and all of a sudden there it is: a tight, orderly cluster of fifty or more trailers—pink and brown and turquoise blue and yellow, aluminum roofs gleaming in the sun, television aerials like a silvery web floating above them; and not a sign of transition between them and the surrounding landscape. I remember in particular a camp situated in a wild and beautiful canyon of dark red rock in Monument Valley. No more perfect setting could have been devised for those simple, cleancut forms, geometrically arranged, with their brilliant colors.[74]

In the 1960s *Landscape* continued its broad coverage. The editor noted that "the plight of that human portion of nature which lives in cities" seemed rather more urgent that other portions,[75] and the magazine reflected this concern in various ways. In 1966 Robert B. Riley, an Albuquerque architect and professor, became his assistant editor with responsibility for "architecture and urban affairs."[76] The dynamics of the contemporary American landscape received special attention,[77] and Jackson continued to bring his unorthodox perspective to bear upon such common features as suburbs, motels, and automobiles.[78] It was the decade of "urban renewal" and "beautification," of the emergence of popular and political attention to the "environment" and the "preservation" of historic and aesthetic buildings and areas. But all of these movements were bedevilled by the lack of well-developed and generally accepted basic principles for the evaluation of landscapes. Jackson devoted much of his magazine to consideration of such matters. Asserting forthrightly that "we have got to find new criteria for the worth of a human landscape,"[79] he set forth his own views in a miscellany of essays, editorials, and brief comments, directly challenging many common and professional attitudes.

He readily agreed that it would be nice to have more beauty in our landscapes, but

> Of all the reasons for perserving a fragment of the landscape . . . the aesthetic is surely the poorest one. And the fact that many still think it the best reason of all derives from a point of view which should be discarded as fast as possible.[80]

He welcomed the burgeoning literature, the big glossy books, awakening us to the visual delight to be found in the forms and textures of our cities, but

> Building watching, like bird watching, is a rewarding hobby. Both can refresh the person who practices them. Both can drive away apathy and ennui and reawaken a sense of vitality in the watcher, by underscoring the beauty to be found in everyday life by one who will search for it. But building watching is a very small and isolated part of the urban experience. It bears the same relationship to living in a city as bird watching does to working a farm or a ranch.[81]

He insisted that it is not a denial of the importance of landscape beauty to believe that

> the blight which is overtaking our country derives not so much from the neglect of esthetic consideration as from the neglect of those factors which make a landscape an agreeable and profitable place to live and work. For the landscape is an essential adjunct, physical as well as spiritual, to the proper development of any individual or society, almost as essential a part of life as the dwelling. Being able to look at it is not enough. It is supposed to serve not one but many purposes, most of them prosaic and far from decorative.[82]

And so if too much of America is "a mess, a garish, jerrybuilt, neon-lighted mess that grows worse each year," simply to label it as 'ugly" and cry for a cleanup is too superficial, "the esthetic judgment is too easy":

> When we try to understand the landscape in living terms, in terms of its viability, our verdicts are not so cocksure. The sordid little businesses which line our highways represent in many cases a last-ditch stand in the field of individual enterprise; tear them down and we have scored a triumph of de-uglification—but at what cost?[83]

"To understand the landscape in living terms"—that is the key, the basic principle. But what does it mean? It means just what he called for in 1956: "to experience the landscape in terms of its inhabitants"; it means

> abandoning the spectator stance and . . . asking ourselves how any man would fare who had to live in it. What chances (for instance) does the landscape offer for making a living? What chances does it offer for freedom of choice of action? What chances for meaningful relationships with other men and with the landscape itself? What chances for individual fulfillment and for social change?[84]

Landscape evaluation must begin with people, and thus any defini-tion of "landscape beauty" must incorporate "a new social dimension": That will require a difficult and humbling change, he argued, because few writers and critics prominent in the discussions of what is wrong with the American landscape have had much to do with its more commonplace workaday aspects, and too many design specialists insisted on overestimat-ing the significance of physical design to human welfare ("architecture is a background to life, but it is not life itself"). Furthermore this principle of dealing with the landscape "in living terms" has an essential corollary: people must be encouraged to look at their surroundings themselves and be given the intellectual and physical tools to do much more of the shaping of their own environments. Such an approach would produce few dramatic or immediate changes, but it would harness the design professions more closely with social needs and be "more compatible with a just democratic society."[85]

    In the spring of 1968, after seventeen years, Jackson gave up *Land-scape*, turning it over to a new editor and publisher, Blair Boyd of Berkeley, California.[86] In a brief and graceful postscript he noted the great change in attitude in the American public during these years, the much greater attention being given to the environment, and he took satisfaction in the fact that year after year *Landscape* had sought and encouraged writers who had something fresh to say about such matters. But he claimed no great credit for this awakening and he recognized that the focus of this new American concern was really divergent from his own. For once Americans' eyes were opened, it was characteristic that they would see their landscapes as "a set of problems to be solved." As editor he had become engulfed by a literature calling for action. But, as Jackson reminded his readers one last time, he had all along been interested in a more humanistic exploration, to see and ponder the landscape "as a symbol of social and religious beliefs and to try to understand blunders as well as triumps as expressions of a persistent desire to make the earth over in the image of some heaven."[87]

Seventeen years was a long time for one man to edit and publish (and finance) a magazine. It was an ever heavier burden, and now as he felt pressures upon *Landscape* to repond to this upsurge of environmental interests, he was ready to let someone else carry it. Furthermore he now had other means of expressing his ideas. In the early 1960s he had been invited to give a lecture at the University of California and this was soon followed by a series of lectures and then a formal course on the history of the American landscape. In the late 1960s he was invited to offer a similar course at Harvard.[88] Thus for a decade his life was seasonally divided among three locations: fall term in Cambridge, winter quarter in Berkeley, spring and summer in New Mexico. These university appointments gave him the chance and the stimulus to adapt and develop his ideas with reference to the whole span of Euro-American history, and since the early colonists came steeped in concepts about the environment, he was led to probe deeply into European antecedents.[89] Freed of the magazine, he could now undertake larger studies. His first book, *American Space, The Centennial Years 1865-1876*, published in 1972, was a study of what he believed to be a critical decade of landscape change. In 1977-78 he retired from both teaching posts, returned to his house and six acres near Santa Fe, and began to work full time on the preparation of a general history of the American Landscape.

Landscape is an impressive exhibit of a remarkable man. It is unique amongst the array of magazines and journals concerned in some way with its general topic. It created a focus where none existed. Begun as a magazine of "human geography," Jackson's successor can now authentically advertise it as a magazine of "human geography, architecture, city and regional planning, landscape architecture, urban and rural history, conservation, historic

John Brinckerhoff Jackson in the yard of his home near Santa Fe, October 1978. (J.B. Jackson)

preservation, environmental design and history, transportation and travel, landscape and the fine arts."[90] Each of these topics has its special literature: no other magazine binds them together.[91] Begun with a hope of appealing to the "intelligent layman," it gathered a readership from many of the best minds in all those fields. It has been one of the most remarkably successful of the "little magazines," those journals of ideas which serve as forums and repositories for essays of the highest quality under the guiding genius of a creative editor. Like that of its more famous counterparts in the literary world, the significance of *Landscape* must be measured by who read it rather than how many—subscriptions never got much above 3000. Anyone with more than a marginal acquaintance with some aspect of landscape studies can thumb through those first seventeen volumes and recognize dozens of writers of national and international reputation. It would be invidious to try to select a few for illustration; it may not be inappropriate to mention one and regard his testimony:

Landscape has made a unique contribution to the thought of our time. . . . No other magazine, with such modest resources, has done so much to focus attention to this major aspect of man's existence.

Lewis Mumford[92]

To assess the significance of *Landscape* to its diverse tributary fields would be a complex task beyond the bounds of this survey, but something should be noted of its relationship with the field it purported to represent at the outset. Jackson's concepts of human geography were deeply rooted in French ideas and practice.[93] This distinctive French school was not unknown to American geographers, but its influence had been relatively minor and certainly little of its deeply humane interpretative spirit had ever been applied to the study of American localities. The term "landscape," however, was widely employed by American geographers, albeit with a good deal of variety and ambiguity. There had been attempts to define geography in terms of landscape analysis, drawing strongly upon German concepts, but by 1951 this had essentially waned.[94] The most famous American formulation had been made by Carl O. Sauer in the 1920s and under his influence a deeply historical and cultural approach to landscape and ecology had become an important though never dominant school of thought in America.[95]

In early issues of *Landscape* Jackson gave a good deal of attention to geography as a professional field, assessing its potentials, prescribing its needs, and declaiming his hopes for explorations of its great themes by scientists and laymen alike.[96] His invitation to give serious attention to rural culture and to learn from close study of the Spanish American landscape soon brought a response from Sauerian geographers, and these together with a growing and diversifying number of other cultural geographers have been a mainstay of the magazine ever since. Nevertheless, despite close affinities at the outset and a long fruitful relationship, Jackson's approach was different from that of the Berkeley School, and as he began to give much greater attention to urban topics and the contemporary American scene, this became more apparent. For ten years he looked for evidence that the mainstream of American geography would give significant attention to the kind of interpretive philosophical exploration of man's ordering of the world he had persistently urged. Failing to find it, the subtitle "Magazine of Human Geography" was in 1961 dropped without comment.

"Landscape." The word could now stand alone on the cover and masthead because Jackson had given it meaning; it now stood for his intellectual as well as his entreprenurial achievement. The meaning with which he clothed the word was laced all through those seventeen volumes, articulated and elaborated issue by issue with reference in total to a great diversity of specific elements and problems, but never, I think, set forth fully in anything like axiomatic fashion. The following is my own distillation from a reading of all of Jackson's publications and manuscripts available to me.

1. The idea of landscape is anchored upon *human life;* "the true and lasting meaning of the word landscape: not something to look at but to live in; and not alone but with other people." Ultimately we study the landscape not as an artifact or work of art, but "for what it tells us about the creator or artist."[97]

2. Landscape is a *unity,* a wholeness, an integration, of community and environment; man is ever part of nature, and the city is basically no less involved than the countryside. The dichotomy of Man and Nature is "a 19th Century aberration and in time it will pass."[98]

3. Therefore we must always seek "to understand the landscape in *living terms,*" "in terms of its inhabitants;" judgments of landscape quality must begin by assessing it "as a place for living and working" and proceed toward a conclusion based on how "productive" it is for the needs of the whole man— biological, social, sensual, spiritual.[99]

4. Just as the elementary unit of mankind is the person, the elementary unit in the landscape is the *individual dwelling,* "the oldest and by far the most significant" man-made element in the landscape. Thus in the study of landscape, "first comes the house," for it is the microcosm, the "most reliable indication of [man's] essential identity." This "ordering of man's most intimate world" is the prototype of how he orders his larger world.[100] Landscape study moves outward from the dwelling also because other basic elements are related to it, both functionally and historically: "The present day farmhouse, surrounded by barns and stables and storehouses and sheds, located near the church, the school, the community center, and the urban residence, is actually the symbol of a parent surrounded by her offspring."[101]

5. To understand the landscape in living terms requires primary attention to the *vernacular,* to the environments of the workaday world. The motel, the franchised fast-food stop, and the contemporary American house seeking to accommodate new mobile and recreational lifestyles are as authentic examples of what vernacular means as the dwelling of a Pueblo Indian or Greek peasant.[102]

6. In the broadest view, all landscapes are *symbolic,* every "landscape is a reflection of the society which first brought it into being and continues to

inhabit it," and, ultimately landscapes represent a striving to achieve a spiritual goal; they are "expressions of a persistent desire to make the earth over in the image of some heaven."[103]

7. And, inevitably, landscapes are ever undergoing *change:* "There is no such thing as a genuinely static human landscape," and because landscape is a reflection of society, if we wish to change the landscape for the better we will have to change the society which created it.[104]

Numerous corollaries and subordinate concepts could be added, but these, it seems to me, define the central ground. Much of this is explicit in his earliest writings and reflects a variety of influences long before he ever conceived of *Landscape*,[105] but all of it was shaped in the long dialectical relationship between the creator and his child: out of the need to assess and react to hundreds of manuscripts submitted, and the need to survey, select, and comment upon the continuous outpouring of literature in several languages. It was a unique position. Jackson was in close touch with half a dozen fields and fettered to none. He was interacting with some of the best minds in several disciplines and guilds but remained himself the intelligent layman, independent of all the institutional and professional pressures and rigidities. There were of course special pressures inherent in the role of publisher and editor. We can see him influenced by the trends of the times and the need to respond to popular topics and important statements of the day, but ultimately he had a wide freedom to select and to speak his own mind.

Furthermore, he claimed the freedom of the layman and of the essayist, and these are basic to the power of his work. The naive, fresh look, unordered by orthodoxy; the clean prose, unsullied by jargon; the conversational tone, unstrictured by analytical forms; the modest, often diffident, manner, unpretentious of special skill or authority—these features, by their very contrast with the prevailing styles of professional writing, are fundamental to the attractiveness of Jackson's work. But therein lie the limitations also. All is assertion and argument, nothing is documented or formally demonstrated; much is observed, nothing is measured. Jackson is a stimulating thinker, he is not a professional scholar. His writings are never supported by the usual research apparatus. There are no footnotes in *Landscape*—and we would not expect nor want them there; but there are none in *American Space* either, even though it is a detailed book-length study of a ten-year period in American history.[106]

None of this vitiates his work, it only underscores its special character. Jackson never purported to be a research scholar, and he openly assumed the role of the speculative intuitive interpreter reaching beyond the usual realms of science. It does mean that much of his most interesting work remains tentative, contingent upon formal and detailed investigation or experiment. An example of this feature is explicit in the first volume of *Landscape*. In the second issue, Jackson had written an essay on the historical and symbolic significance of the American front yard, tracing its roots deep into the cultural history of Northwest Europe. A reader challenged Jackson's interpretation, arguing that it was simply invalid because there was no such historical continuity, that in fact the American front yard was a mid-nineteenth century invention. Jackson printed the letter but offered no rebuttal.[107] That question and many of his other interpretations must remain open until we have much more research on the making of the American landscape.

Similarly, Jackson's premises about planning are attractive and his long running critique of the design professions usually seems persuasive, but taken as a whole there remain many uncertainties in his position. Aside from problems of practicality in his insistence on a more democratic approach to planning, his view of the relationships among the ecological, social, and aesthetic aspects remain unclear as a guide to specific planning, and it becomes difficult to define just what he means by aesthetics and how he incorporates it into landscape appreciation. If he scorns the "beautificationists" as superficial and asserts that "ugly" is *"nothing but a euphemism"* for "poverty, disease, greed, laziness, corruption,"[108] he does not really tell us how to combat those kinds of "uglification." In one rather powerful essay on the plight of small towns and small cities, he focused very directly on poverty and political inertia as the root causes and he suggested some fairly specific topics to which planners and politicians should be giving attention, but his main theme was the need for an end to the neglect of such places and for a more sensitive mode of planning.[109] Jackson is certainly aware that the worst landscape problems are bound up with the political and economic system, but it also seems clear that he is not basically interested in planning, and certainly not in politics. He is interested in attitudes and interpretations. He spent seventeen years trying to change many common attitudes, but there is no evidence in his magazine that he really wanted to get directly involved in actually changing the landscape. There is some significance in the fact that he turned away from *Landscape* just as the political and social critique of American society, including what it was doing to its landscapes, was becoming very shrill and intense. That critique made good use of a basic Jacksonian premise: that landscape mirrors society. The American people were opening their eyes to their surroundings, as Jackson wanted them to do, but they were shocked and outraged at what they saw, and he was not.

It was not a matter of his liking everything he saw, it was a matter of his seeing everything from a different point of view—in part he would insist, from a more *American* point of view. He had repeatedly argued that the root of much of our difficulty in developing effective approaches to design and planning, preservation and appreciation, was the fact that the critics and spokesmen

> have almost always traveled and studied in Europe, and have been properly enthusiastic about what they experienced there. To them it seems logical that we should undertake to transform America—insofar as this is possible—in the Western European image. If only the public could be aroused. . . . [110]

But the beauty of those admirable European landscapes is the expression "of a stable and well-ordered society respectful of its past and confident that its values will be honored by future generations." Such landscapes can neither be expected nor imposed in a country "where land and buildings are increasingly thought of in speculative terms, where families move on the average of once every five years, where whatever is old is obsolete, and whatever is obsolete is discarded.. . ."[111]

"If the landscape is to be truly human it ought to reflect the kind of people we are."[112] For Americans that means "recognizing and accepting our national landscape for what it is: something very different from Europe."[113] Because Americans so value change, our landscape "*must* be fluid and capable of sudden change."[114] We must learn to see that "most American communities, large or small ... are utilities intended to grow and change—and if need be disappear." In its more extreme forms change might be deleterious; but at their best, our cities "have a remarkable flexibility and versatility and capacity for assimilation" and we must learn to judge them "not as works of art but as a kind of makeshift environment for highly regarded social institutions; as such they appear to be adjusted to very basic traits in the national character."[115]

Whereas critics of America saw in our landscapes alarming exhibits of social problems, Jackson saw something more, something deep, fundamental. For if landscape is to be read as a reflection of life, if one lay oneself open to this "immense and tawdry and beautiful American countryside,"[116] if one looked unflinchingly at the rich variety of the workaday world, at the ordinary, the genuinely vernacular landscapes of modern America, one could not but be impressed with the *immense vitality* of American life. Moreover all the seeming chaos, the confusion, clutter, and dereliction so rampantly apparent in the American landscape could itself be read as a sign not of what was wrong, but more fundamentally of the search for what was right, the search for a new concept of how to create heaven on earth. For beauty is not something pretty to be preserved: it is something that inheres in the congruence between the landscape and the strivings of the spirit. In the dawn of our national life just such a striving had initiated a landscape of classic beauty, the spatial order of the Jeffersonian Agrarian Utopia, a concept no longer viable but the residual framework for much of rural and small town America. Later on, the Romantic view of Man and Nature left its mark upon our domestic architecture and upon our city parks and cemeteries, suburbs and campuses. We have turned away from those visions but we have not as yet articulated coherently and persuasively a successor; and so we seem to have, for the moment, lost "the capacity—or the temerity—to construct Utopias."[117]

But the search goes on and the landscape reflects these very uncertainties and strivings. An "existential landscape" is already taking form around us, he wrote, "without absolutes, without prototypes, devoted to change and mobility and the free confrontation of men. . . . It has vitality but it is neither physically beautiful nor socially just."[118] We live in a time of ever greater emphasis upon individual pleasures and emotions, upon inner-directed movements and the heightening of personal experiences, upon the search for new levels of consciousness. At their deepest these explorations as well as the probings of the scientist into the invisible aspects of creation, can be read as part of the search for a new vision of order, and that order will be displayed in a new landscape. Jackson, with his characteristic openness to whatever the fundamentals of living might do to the landscape, recently expressed his great interest in its emergence. At the moment, he said, he could discern no more than a few fragments, but if called upon to cite an archetypal representation of this new man and new landscape, of this emerging scene

> dedicated as it seems to be to danger and uncertainty and movement, I would choose the man who is discovering a new kind of peak experience—not in worship or work, but in interaction with environment—the man on the skateboard, in a kind of ecstasy of mobility, physical grace, and awareness.[119]

Conclusion

William G. Hoskins and John Brinckerhoff Jackson are approximately of an age, products of the same era, influential upon landscape studies over almost exactly the same time. They are clearly alike in being gentlemen of high intelligence, wide learning, acute observation, and literary skill: qualities which have been basic to their cultivation of landscape study as a humane art. In each case, behind the quietly courteous manner, one senses a strong personality, a man of ideas and convictions, of energy and spirit. Neither has been an orthodox professional. Jackson was of course much the freer of the two, always the layman, self-supporting, never a regular faculty member. Hoskins spent most of his life in the academic world, but early on as a student and a teacher in university colleges he learned his research skills yet escaped any heavy doctrinal influence ("there is much to be said for being self-taught, for never having sat at the feet of some immense authority laying down the law in unforgettable terms").[120] He soon created his own field and department, and eventually early retirement allowed him to explore new forms of expression. Such freedoms were important to their efforts to reach out "in a less didactic manner" (JBJ) to a general public in order "to communicate to others the pleasure it gives us" (WGH) in taking "the landscape as an object of contemplation in both senses of the world—seeing and pondering" (JBJ). Both have unabashedly written about that which they love.[121]

Both men have sought "the logic that lies behind the beautiful whole," but they have seen a different beauty and have searched with different tools. In a broad sense both have an historical view, but in practice they are utterly different. Hoskins is the research scholar searching for the truth in the details of this particular house, this lane, this hedge, by means of the most meticulous work in field and archive; Jackson is the intuitive thinker seeking the truth in this kind of house, this repetitive pattern of farm and town, this overall design of the landscape as read from a rapid reconnaissance. The one is engaged in a skilled detective work of exact dating from scraps of evidence; the other in an interpretive work of general meaning from a keen sense of culture history.

In assessing the character of their work one cannot but begin to think of each as representative of a common stereotype of his national culture. The one deeply rooted in ancestral ground, absorbed in local study, exploring the countryside on foot and bicycle (he never learned to drive an automobile), drawing strength from the feel of history through his very boots

("How could one bear to live in a country with only a hundred or two hundred years underfoot, above all bear to die in it?"[122]); looking backward to better times, eyes fixed on remnants of this green and pleasant land, crying for their preservation against all the "horrors" of modernity;[123] empirical in research, uncomfortable with generalizations, eschewing abstractions, finding deep satisfaction in a kind of learning which need have "no practical or theoretical importance whatsoever."[124]

The other choosing to live in a region thousands of miles and utterly different from that of his youth and family, cosmopolitan and urbane, but with a strong affection for rural and small town communities and a desire to help improve them; an inveterate traveler, crossing the continent and the ocean time and again, doing field work from car, truck, plane, and motorcycle; receptive to the common scene ("I have been plagued by an urge to understand the contemporary landscape—to understand and love it regardless of its unsightliness,"[125]), open to change,[126] attracted by new experiences, looking toward the future ("how I wish I could live to see it"); appreciative of European qualities but in love with the "varied and exhuberant beauty" of his own country ("It is impossible after traveling through the American landscape from city to town to village, not to feel a love for it, and finally a pride of possession"[127]); ready to generalize at a continental scale, searching the ever "growing and shifting design" on the land for the signs of Utopia.

We should be cautious here, for such comparisons can be pushed all too easily into caricature, both of the persons and of the nations. They are useful however in suggesting that there is indeed a deep cultural logic underlying these very different approaches and attitudes toward landscape appreciation. Serious examination of that relationship would require more extensive biographical as well as cultural inquiry into these two cases. Hoskins's obsession with the past and Jackson's eagerness for the future are surely expressions of temperament as well as nationality. However, to whatever degree we may accept these approaches as emblematic of their particular cultures we need not assume that they are culture-bound. The Americans and the English can learn from one another in landscape study as they have in countless other topics in their enduring, shifting transatlantic dialogue.

In his stimulating inquiry into this lengthy Anglo-American interaction, Stephen Spender has noted how much of it can be summarized under the idea that "Americans fear the European past; Europeans fear the American future."[128] The characterization is pertinent: Jackson repeatedly asserting the dangers of importing a European aesthetic and sentimentality; Hoskins repeatedly denouncing all that is "modern"—by obvious implication all that material culture for which America is world famous. In extreme form Spender observes that such fears have often led to a rhetoric of recrimination in which the "Europeans [are] left with their ruins, Americans with their hardware and glittering junk."[129] But if we wish to gain from this dialogue we will have to stress complements rather than contrasts.

The most obvious possibility is to try to bind together the contrasting emphases upon the past and upon the present and future, to work toward an appreciation of the landscape as a continuum of change. The Hoskins approach must be rescued from its determined preoccupation with the past. We can indulge and enjoy Hoskins himself in the vigorous assertion of his taste, but we cannot allow the twentieth century to be excluded from serious investigation of "the making of the landscape." Here Jackson points the way in his insistence on looking the modern scene squarely in the face; and his admonition is not simply for us to be comprehensive and tolerant, but to see the ordinary landscapes of the automobile, mobile home, supermarket, and shopping center as legitimately "vernacular"—that is, native to the area, but area now defined more at the national than the local scale. The Englishman may sense "the spectre of Americanization" in much of the modern he sees, but whatever ideas or elements may actually have been imported they can never result in an American landscape; the selections, settings, and adaptations remain peculiarly English. Underlying this positive approach is Jackson's tolerance of change, and underlying that, of course, is his basic principle of evaluating landscapes in terms of *life*. This insistence upon a social as well as an aesthetic dimension may well alter our appreciation, but can save us from nostalgia and sentimentality. In more practical terms it points toward an emphasis upon the search for a truly humane conservation rather than rigid preservation of patches of history and beauty.

On the other hand, it seems equally clear that Jacksonian interpretation requires the Hoskins approach for substantiation. We will need to know much more about the detailed making of the landscape before we can be very confident of its meaning in the past, present, or future. In America, especially, that will require not only the kind of meticulous local study of the past and the particular so well exemplified in the work of Hoskins and his associates, but special attention to the source and diffusion of ideas and elements in the creation of a very strong national component in local landscapes. The skills of a Hoskins do not require a thousand years of history to make deciphering worthwhile. Much of the American vernacular landscape is also written in a code complex enough to warrant their use.

Hoskins can also teach us more than how to do history, he can teach us much about landscape viewing as a humane art. His likening the landscape to a symphony is the most telling metaphor for it points to an aesthetic of unlimited challenge and enrichment. Hoskins is attuned to English compositions, but the concept is applicable anywhere. Jackson's own aesthetic sensibilities show through in much of his writings, but he seems to reject the asesthetic as a fundamental component in landscape appreciation because it appears to conflict with his basic premise that landscapes must be regarded first of all in terms of living rather than looking. We may accept his stance as a warning against the insidious danger of treating people in a landscape either as picturesque elements or irritating intrusions, but his implication that the aesthetic is relatively unimportant seems all too American. Landscape appreciation must be grounded upon life, but the ideal is to have looking accepted as a routine part of living, to have aesthetic consciousness and satisfaction regarded as an essential component of a healthy and happy life *in* any landscape.

These comments suggest a direction but they barely hint at the complexities and nuances inherent in the topic. It would be fatuous to declare that the best of the Jacksonian view could simply be applied in England or that of Hoskins in America, for these societies have critically different attitudes toward the past and the future and the landscape. As an American, I would welcome a strong injection of a Hoskins sense of history and aesthetics into our landscape studies, but I am aware that it could not have widespread effect without an alteration of common cultural values. That need not deter the attempt, it simply warns that we will have to see such studies themselves as instruments for trying to effect such alterations. And I believe that I have crossed the Atlantic enough times to realize that a Jacksonian verve for the existential landscape of today and his search for the signs of utopian design are simply not directly appropriate to the English scene. Change cannot have the same meaning in both societies; nevertheless there is much in his premises which could help the English cope with the ongoing making of their own landscape.

Reading the landscape is a humane art, unrestricted to any profession, unbounded by any field, unlimited in its challenges and pleasures. These two men are surely among the best exhibits of that truth. They are important not only for what they have said, but how they have said it; they have spoken their minds and expressed their feelings. They have created a literature in which we become engaged by their persons as well as by their topics, and each is the kind of person to whom we can feel immensely indebted without feeling bound, for neither has sought to become "an immense authority laying down the law." They have taught by the spirit rather than by the letter, and it may be well to remind ourselves, in conclusion, that there are still other ways of reading the landscape, other "logics" that lie behind the beautiful whole. There is the logic, for example, of political systems with their concepts of law and property and their discriminatory powers over social actions and aspirations. Increasingly strident voices on either side of the Atlantic are implying that this logic is also part of the making and the meaning of the English and the American landscapes. In cultivating our landscape sensibilities we should attune ourselves to hear other harmonies and dissonances in further interpretations of this grand earthy composition.

Notes

1. *Landscape* 1 (Spring 1951): 5. (The magazine will hereafter be cited as *L*)
2. *The Making of the English Landscape* (London: Hodder & Stoughton, 1955), p. 14. (Hereafter cited as *MEL*)
3. *MEL*, p. 19.
4. *English Landscapes* (London: British Broadcasting Corporation, 1973), p. 5.
5. Ibid.
6. In the Preface of *Leicestershire, An Illustrated Essay on the History of the Landscape* (London: Hodder & Stoughton, 1957), Hoskins mentioned "I owe much to my old friend Mr. F.L. Attenborough, who took many of the photographs in this book on our jovial expeditions together in past years." Seventeen plates in *MEL* were contributed by Attenborough; see p. viii.
7. Hoskins recalls that he hesitated over this invitation, for he was determined not to give up his commitments to local history and the landscape, and he accepted only after being reassured that the reputation of the ancient university for instructional freedom was unimpaired: "My dear chap," he was told, "you can lecture in Chinese if you want."
8. First given in 1955 as three talks on the Third Programme, BBC, and published in *Provincial England, Essays in Social and Economic History* (London: Macmillan 1963), pp. 209–29.
9. These included the volume on Leicestershire by Hoskins, cited above, and W.G.V. Balchin, *Cornwall;* H.P.R. Finberg, *Gloucestershire;* Roy Millward, *Lancashire;* all published by Hodder & Stoughton.
10. G.E. Fussell, *The English Historical Review* 71 (April 1956): 327. E.G.R. Taylor gave the book a moderately favorable review in *The Geographical Journal* 121 (December 1955): 511–13.

11. H.P.R. Finberg, an early collaborator with Hoskins on Devonshire studies, succeeded him as head of the department in 1952 and became the first professor when the chair was established in 1964.

12. (London: Penguin Books, 1970). The book is identical in content except for an updating of the bibliography, which was shifted from chapter endings to the back of the book.

13. Roy Millward, Reader in Geography at the University of Leicester and author of the Lancashire volume in the original series, has joined Hoskins as coeditor of the new series, all published by Hodder & Stoughton: Christopher Taylor, *The Cambridge Landscape* and *Dorset;* John Steane, *The Northamptonshire Landscape;* Robert Newton, *The Northumberland Landscape;* Frank Emery, *The Oxfordshire Landscape;* Trevor Rowley, *The Shropshire Landscape;* Moelwyn Williams, *The South Wales Landscape;* Norman Scarfe, *The Suffolk Landscape;* Peter Brandon, *The Sussex Landscape;* Arthur Raistrick, *West Riding of Yorkshire;* David Pallister, *Staffordshire;* Lionel Muncy, *Hertfordshire.* Finberg's *The Gloucestershire Landscape* has also been republished in the new format.

14. London: Hodder and Stoughton, 1977. This book is identical in content with the 1970 Pelican edition, except for a new introduction in which Hoskins noted some of the pertinent literature which had appeared since 1955, commented on new data relating to the age of and continuity in the English Landscape, and stated his reasons for not revising his 22-year-old book: "There is so much we still do not know, so much work in progress, that a revision is still premature. Moreover . . . there is continual change in the English landscape, at an accelerating pace during the past generation or so."

15. The regional films are: Series I: *Ancient Dorset, Conquest of the Mountains* (Lake District), *Marsh and Sea* (Norfolk), *Landscape of Peace and War* (Romney Marsh & the Weald), *The Deserted Midlands, The Black Country;* Series II: *Behind the Scenery* (Cornwall), *The Fox and the Covert* (hunting landscapes of East Midlands), *No Stone Unturned* (industrial landscapes of Derbyshire), *Brecklands and Broads* (Norfolk), *The Frontier* (Northumberland), *Haunts of Ancient Peace* (Devon).

16. See Note 4.

17. *Rutland, A Shell Guide* (London: Faber & Faber, 1963). *Leicestershire* (London: Faber & Faber, 1963). It is interesting to note that Part Two of his monumental volume on Devon is also a gazetteer of places, not unlike a Shell Guide. The first edition was published in 1954 and went through four printings; a new edition was published by David & Charles, Newton Abbot, 1972.

18. As examples of literature clearly aimed at a broad popular readership see John Patten (historical geographer at Oxford), "The Shaping of England's Landscape," *Observer Magazine,* 21 April 1974, pp. 40–50, with its imaginative panel of pictorial diagrams illustrating landscape change through five eras; *The Changing Face of Britain* (Harmondsworth, Middlesex: Kestrel Books, 1974) reissued as a Paladin paperback (St. Albans: Granada, 1977) by the versatile writer Edward Hyams, which seems to be almost certainly modeled on the idea of Hoskins's book but much more generalized, and rather different in emphasis.

19. Some of his books and collections are: *Industry, Trade and People in Exeter,*

1688–1800, 1938 (new edition, 1968); *The Heritage of Leicestershire*, 1946; *Midland England*, 1949; *Devonshire Studies* (with H.P.R. Finberg), 1952; *Devon*, 1954; *The Midland Peasant*, 1957; *Provincial England*, 1963; *Fieldwork in Local History*, 1967; *History from the Farm*, 1970; *The Age of Plunder*, 1976. It is interesting to note that a recent survey by historians recognized him as a "master in the field of local history, with "many distinguished contributions," but gave no specific notice of his landscape studies; see, *Changing Views on British History*, ed. Elizabeth Chapin Farber, (Cambridge: Harvard University Press, 1966), Hoskins ref. pp. 135–36. He was on the Honours List in 1971, a C.B.E. for "services to local history," which presumably include all of his landscape studies. In 1973 he was made an Honourary Fellow of the Royal Institute of British Architects in specific recognition of his work on the English landscape.

20. *MEL*, p. 18.
21. *Fieldwork in Local History* (London: Faber and Faber, 1967), p. 171.
22. He has said this in various ways in numerous places. I have taken this quotation from his very favorable review of K.P. Witney, "The Jutish Forest," *Times Literary Supplement*, 17 June 1977, p. 736.
23. *MEL*, p. 119.
24. The quotations are from his Inaugural Lecture, *English Local History, The Past and the Future* (Leicester: Leicester University Press, 1966), pp. 20, 21, 22. In his further discussions of the fundamental value of local history in *Fieldwork in Local History* he has cited Marc Bloch, *The Historian's Craft*, in his support. An American might readily think of James Malin, an unorthodox historian and stout defender of the same theme; cf. "On the Nature of Local History," *Wisconsin Magazine of History*, Summer 1957, pp. 227–30.
25. Arnold J. Toynbee, *A Study of History*, vol. 1 (London: Oxford University Press, 1934), pp. 17–50.
26. "The Rediscovery of England," *Provincial England*, p. 219.
27. *English Landscapes*, p. 6.
28. "The Rediscovery of England," *Provincial England*, p. 228.
29. In the introduction to the new edition of *The Making of the English Landscape*, 1977, pp. 12–13.
30. "The Rediscovery of England," *Provincial England*, pp. 229, 228.
31. *MEL*, p. 231.
32. *MEL*, p. 232.
33. *Rutland*, p. 8.
34. *Leicestershire, An Illustrated Essay*, p. 125.
35. *Rutland*, p. 7.
36. *Leicestershire, An Illustrated Essay*, p. 133.
37. Cf. Edward Hyams, *The Changing Face of Britain*, which is surely a derivative from Hoskins, but which has a chapter, "The Advent of the Car and Electricity," and another on the vast rebuilding, "After the Bombing." Hyams also has strong personal views but he is quite ready to look the twentieth century in the face and to assess the landscape as in some way related to social needs.
38. *MEL*, p. 231.

39. *MEL*, pp. 231–32.
40. *MEL*, p. 109.
41. *English Local History*, p. 21. He is relating how he resisted calling his field Micro-History, inventing a vocabulary of esoteric jargon, and making use of the computer: "A new title for one's subject, a glossary of jargon, and a computer, and one has the most lethal combination for academic advancement conceivable. One would then qualify to work in some shiny academic palace." In such derision we presumably have another reflection of the discomforts of an unorthodox professor with the bureaucracy and guilds of university life.
42. *Leicestershire, An Illustrated Essay*, p. 131, 130–31.
43. *Leicestershire, An Illustrated Essay*, p. 84.
44. *Rutland*, pp. 15–16.
45. *L* 1 (Spring 1951): 41.
46. *L* 1 (Spring 1951): 16–24.
47. Published in Paris, 1948–49. He was also well acquainted with other French, German, and Swiss magazines in geography and related fields.
48. *L* 1 (Spring, 1951): 4–5.
49. "1951–1968: Postscript,' *L* 18 (Winter 1969): 1.
50. The essays are "Ghosts at the Door," *L* 1 (Autumn 1951): 4–9, and "The Almost Perfect Town," *L* 2 (Spring 1952): 2–8.
51. "What We Want," *L* 1 (Winter 1952): 2–5. In an earlier issue he mentioned his own direct response to this problem: the establishment of a *Landscape* Prize to be awarded annually in the schools of Santa Fe County for the best Spanish compositions on some theme dealing with people and their localities. Jackson set the topic each year and conducted this contest until 1977.
52. "Human, All Too Human Geography," *L* 2 (Autumn 1952): 7.
53. "The Westward Moving House: Three American Houses and The People Who Lived in Them," *L* 2 (Spring 1953): 8–21. See also his essay "Pueblo Architecture and Our Own," *L* 3 (Winter 1953–54): 20–25.
54. Including Jackson's own essays which, in addition to those cited above, included "High Plains Country, A Sketch of the Geography, Physical and Human, of Union County, N.M.," *L* 3 (Spring 1954): 11–22.
55. *L* 4 (Winter 1954–55): 2.
56. *L* 6 (Autumn 1956): 2–5.
57. *L* 2 (Spring 1953): 34–35.
58. *L* 3 (Summer 1953): 28–29. See also his comments on the Bauhaus, *L* 3 (Summer 1953) pp. 4–5, and his devastating treatment of "gracious living" in a Mies van der Rohe house, *L* 3 (Spring 1954): 24–25.
59. "Other-Directed Houses," *L* 6 (Winter 1956–57): 29–35.
60. "The Stranger's Path," *L* 7 (Autumn 1957): 11–15.
61. "The Abstract World of the Hot-Rodder," *L* 7 (Winter 1957–58): 22–27.
62. Ibid., pp. 26–27.
63. "Southeast to Turkey," *L* 7 (Spring 1958): 17–22, ref. p. 21.
64. "The Imitation of Nature," *L* 9 (Autumn 1959): 9–12, ref. p. 10; the last sentence here has been transposed from a preceding paragraph.

65. Ibid., p. 12; see also "Our Unexplored Suroundings," *L* 8 (Autumn 1958): 26–28, in which he writes of the need for bioclimatological studies of urban environments.

66. "We have rarely left the region [the Southwest] for long, and never willingly." *L* 10 (Fall 1960): 10.

67. "First Comes the House," *L* 9 (Winter 1959–60): 26–32; ref. pp. 32, 26.

68. *L* 10 (Fall 1960): 1–2.

69. Ibid., p. 2.

70. Ibid.

71. "The Four Corners Country," *L* 10 (1960): 20–26.

72. He noted that "neither the critics nor the defenders of the trailer house have tried to recognize the trailer for what it is," and "typical of the general ignorance" was the failure even to mention the house trailer, home for four million Americans, in *Fortune's* recent survey on coming changes in housing in the 1960s. Ibid., p. 24.

73. Ibid., p. 26.

74. Ibid. p. 24.

75. *L* 13 (Autumn 1963): 3.

76. *L* 15 (Winter 1965–66): 1.

77. E.g., special sections featuring news items and comment on "The Evolving Landscape," "The Evolving Strip," and "The Evolving Highway" were added.

78. On suburbs see his preface to an article by Herbert Gans, *L* 11 (Autumn 1961): 22; on motels see "Designed for Travel," *L* 11 (Spring 1962): 6–8. His one page commentary on the automobile is the epitome of his kind of incisive, illuminating critique. He asserted that a major deterrent to successful urban design has been "the planners' emotional aversion to the automobile [and] their total failure to understand the reasons for its popularity." These reasons go far beyond the functional; they are deeply emotional and especially involve "territoriality," those "structured volumes of psychologically differentiated space" which persons perceive and carry with them. "Put simply, the automobile allows one to travel almost at will anywhere in the public domain while remaining in a completely private world unequivocally defined by physical boundaries." "Auto Territoriality," *L* 17 (Spring 1968): 1–2; see also "The Domestication of the Garage," *L* 20 (Winter 1976): 10–19.

79. "Goodbye to Evolution," *L* 13 (Winter 1963–64): 1.

80. Ibid.

81. "The Building Watchers," *L* 16 (Autumn 1966): 2.

82. "Limited Access," *L* 14 (Autumn 1964): 18–23; the article begins as a critical review of Peter Blake, *God's Own Junkyard*.

83. "The Hazards of Uglitudinizing," *L* 12 (Autumn 1962): 1–2.

84. "Goodbye to Evolution," *L* 13 (Winter 1963–64): 2.

85. "Pretentions and Delusions," *L* 17 (Winter 1967–68): 1–3, a succinct and severe critique of "behavioralism" as the latest fad in the design professions. See also *L* 9 (Spring 1960): 1–2.

86. Boyd is professor of landscape architecture at the University of California. The

sale was consummated for the price of $1. After a considerable hiatus, *Landscape* is again being issued three times a year from Box 7177, Landscape Station, Berkeley, CA 94707.

87. "1951–1968: Postscript," *L* 18 (Winter 1969): 1.
88. His appointments were in each case in the department of Landscape Architecture; for a time at Berkeley he was partly supported by the Department of Geography. He has also been guest lecturer at numerous universities.
89. Thus the first section of his book in preparation on the history of the American landscape is entitled "The First Thousand Years" and begins in medieval Europe.
90. As listed in a broadsheet circulated in 1976.
91. Note also that Marvin Mikesell concluded his article "Landscape" in the *International Encyclopaedia of the Social Sciences* (vol. 8, [New York: Crowell-Collier and Macmillan, 1968] pp. 575–80) by pointing to Jackson's *Landscape*: "Indeed the diverse contributions to this magazine—from architects, ecologists, geographers, planners, and perceptive observers from many other backgrounds—provide an effective illustration of the continued value of landscape as an integrating concept in the social sciences." Jackson has had a similar kind of influence through his teaching. His courses at Harvard and Berkeley and his guest lectures on many campuses brought the significance of formal study of the landscape to the attention of important educators, and especially to those in charge of university training in the design professions. Thus, with his retirement impending, a committee of the Association of Collegiate Schools of Architecture launched a drive for the creation of an endowed national Chair of American Landscape which would be rotated among member schools in order to catalyze work on the topic; see remarks of the editor, *Journal of Architectural Education* 30 (September 1976): 2. When I inquired of this editor just what relationship there might be between this idea and Jackson's work, he replied that the Chair "was inspired by him, wouldn't exist (such as it is) without him, and intends to carry on his work"; letter from D.S.C. Clarke, 12 September 1977.
92. Mumford's letter was one of 28 published in response to Jackson's invitation to "some of our earliest subscribers and supporters" to express their views on the occasion of the tenth anniversary of *Landscape*. *L* 10 (Fall 1960): 3–7.
93. The first issue draws explicitly upon Le Lannou, later issues upon Deffontaines and Sorre, and his general approach is certainly influenced by the traditions of Vidal and Brunhes. For a succinct assessment of these men see Anne Buttimer, *Society and Milieu in the French Geographic Tradition*, Monograph Series, Association of American Geographers (Chicago: Rand McNally, 1971).
94. Richard Hartshorne, *The Nature of Geography*, A.A.G. (Lancaster, Pa. 1939), esp. chap. 5; Preston E. James, *All Possible Worlds, A History of Geographical Ideas* (Indianapolis: Odyssey, 1972), esp. chaps. 8 and 13.
95. Carl O. Sauer, "The Morphology of Landscape," *University of California Publications in Geography*, No. 2, 1925.
96. See especially "Human, All too Human Geography," *L* 2 (Autumn 1952): 2–7; "The Unknown Country," *L* 3 (Autumn 1953): 1; p. 29 of this issue his comments on George Stewart's *U.S. 40;* and *L* 3 (Spring 1954): 3–4.

97. "The Meaning of Landscape," (Lecture given at Rensselaer Polytechnic Institute, 21 November 1977); "Prologue" to "Teaching the Landscape," special issue of the *Journal of Architectural Education*, ed. J.B. Jackson, 30 (September 1976): 1.

98. He once used the term "Gestalt," *L* 16 (Autumn 1966): 2, but he usually avoided such jargon; the quotation is from "The Imitation of Nature," *L* 9 (Autumn 1959): 10.

99. His most impassioned and lengthy statements on this theme are "Limited Access," *L* 14 (Autumn 1964): 18–23 and "To Pity the Plumage and Forget the Dying Bird," *L* 17 (Autumn 1967): 1–4; specific use of the "productive" concept is in *L* 6 (Autumn 1956): 3.

100. The sequence of quotations is from: *L* 1 (Winter 1951–52): 10; *L* 9 (Winter 1959–60): 26; *L* 2 (Autumn 1952): 6 (last two quotations).

101. *L* 1 (Winter 1951–52): 10.

102. Cf. "The Domestication of the Garage" *L* 20 (Winter 1976): 19.

103. For an early extensive statement of this theme see "Human, All Too Human Geography," *L* 2 (Autumn 1952): 5–7.

104. Quotation from "Southeast to Turkey," *L* 7 (Spring 1958): 21.

105. His readings in the European literature previously noted are the most obvious source; Spengler's emphasis upon a spiritually empowered gestalt was also influential, and Jackson's concepts seem closely related to Mumford's persistent "organic" critique of industrial civilization. Jackson recalls the humanist Irving Babbitt as his most influential teacher at Harvard; he also had two "very good" courses in geography from Derwent Whittlesey.

106. Some of his sources are identified directly in the text, some others can be inferred from the brief reading list appended.

107. *L* 1 (Winter 1952): 40.

108. *L* 12 (Autumn 1962): 1–2.

109. "To Pity the Plumage and Forget the Dying Bird," *L* 14 17 (Autumn 1967): 1–4.

110. "Limited Access," *L* (Autumn 1964): 19–20; see also his reviews of Jacquetta Hawkes, *A Land*, *L* 2 (Autumn 1952): 34–35 and of Ian Nairn, *The American Landscape L* 14 (Spring 1965): 35.

111. *L* 14 (Spring 1965): 1.

112. *L* 13 (Winter 1963–64): 2.

113. "Metamorphosis," *Annals*, Association of American Geographers 62 (June 1972): 158.

114. *L* 13 (Winter 1963–64): 1.

115. *L* 14 (Spring 1965): 35.

116. *L* (Autumn 1953): 29.

117. "Jefferson, Thoreau and After, The Life and Death of American Landscapes," *L* 15 (Winter 1965–66): 25–27.

118. Ibid., p. 27.

119. "The End of Landscape." Lecture given in the College of Environmental Design, University of California, Berkeley, 15 March 1977.

120. *Field-Work in Local History*, p. 171.

121. We should note here that both men were interested in the creation of a literature to assist the "intelligent layman" in his travels. Hoskins's Shell guides are
an obvious exhibit. In the first issue of *Landscape* Jackson had a section on
tourism, and his comments on the topic are sprinkled through subsequent volumes. He praised such creative works as George Stewart, *U.S. 40* and Eric
Sloane, *Return to Taos*, and in 1962–63 he published three pocket-size auto tour
guides to Northern New Mexico, *L* 11 (Spring 1962): 1; see also pp. 20–22 "We
are Taken for a Ride,' an indictment of "this monstrous parasite called the
Tourist Industry" and a call to make it an essentially geographic experience. His
little guidebooks are informative but have none of the poetry of a Hoskins, nor
even of his own most evocative essays. Typically Jackson was really more interested in national attitudes and in the behavior of tourists than in creating a
literature for them.

122. *English Landscapes*, p. 6.

123. He would have the whole of Rutland set aside as a preserve if he could, for it is
still:

> . . . a picture of a human, peaceful, slow-moving, pre-industrial England, with
> seemly villages, handsome churches, great arable fields, and barns. One
> would like to think that one day soon at each entrance to this little county,
> beside a glancing willow-fringed stream, there will stand a notice saying,
> Human Conservancy: Abandon the Rat-Race at this Point.

Rutland, p. 7.

124. "The Rediscovery of England," p. 228.

125. Letter to the author, 24 September 1977.

126. And therefore never comfortable with the idea of preservation: "the preserva-
tionist is constitutionally unable to design in terms of changing patterns of use.
He provides us with the experience of extraordinary beauty, he makes it possible for us to step out of the everyday world; but he does nothing to make that
everyday world more liveable." *L* 14 (Autumn 1964): 23.

127. *L* 17 (Autumn 1967): 1.

128. Stephen Spender, *Love-Hate Relations, A Study of Anglo-American Sensibilities*,
(London: Hamish Hamilton, 1974), p. 47.

129. Ibid., p. 5.

The Contributors

and something of the landscapes in their lives

Years of residence, schooling, and travel in France, Switzerland, Austria, and New England whetted JOHN B. JACKSON's interest in landscape, which he put to practical use as a combat intelligence officer in World War II. He has spent much of his life trying to bring a good, old-fashioned Harvard education up-to-date and to interpret the contemporary American landscape as a fulfillment of changes which began two centuries ago. For many years home has been a village near Santa Fe.

PEIRCE LEWIS grew up along the shores of the upper Great Lakes, and he served his apprenticeship as a student of landscape in the wild dune country of western Michigan. Since then, he has lived and studied in South Carolina, Japan, western Washington, the Spanish meseta, the Willamette Valley, French Canada, New Orleans, and central Pennsylvania, where he is now Professor of Geography at Pennsylvania State University. Like many Americans, he is fascinated by what humans have done to the earth, but is fondest of places where they have exercised restraint. It is a minor paradox that he is shamelessly in love with California.

DAVID LOWENTHAL has lived in landscapes of the mind too long to remember much about the real world. In summer he always stays at home; he has, however, passed through all the states and continents but one. Among his favorite retreats are People's Park, Little Big Horn, Bull Run, and the Teutoburger Wald. A reformed New Yorker, he now seeks solace in nineteenth-century London, and serves as Professor of Geography at University College. W.H. Hudson and George Orwell both took credit for inventing Lowenthal, but he is actually the eponymous hero of Clough Williams-Ellis's *Britain and the Beast*. His preferred landscape is Orcadian.

The Palouse Hills and a Pacific Northwest version of Main Street were nurturing landscapes for DONALD W. MEINIG. Subsequently he has spent periods varying from several months to several years in Seattle, Colorado Springs, Salt Lake City, Washington, Adelaide, St. Andrews, Jerusalem, and the Wye Valley; for the last twenty years he has lived contentedly amidst the drumlins of Upstate New York where he is Maxwell Professor of Geography at Syracuse University.

Upstate New York and Lower Manhattan served as initial ground for MARWYN S. SAMUELS. For university studies he moved on to Denver and Seattle, but his mind turned increasingly to Asian scenes, and he has since spent various periods of time in Hong Kong, Taipei, Peking, Shanghai, Manila, San Pablo, Singapore, Soviet Central Asia, Jerusalem, and Doha—in addition to Berkeley and Vancouver. Foremost in his landscape reference points are: China under the *ancien regime;* Manila in the company of his wife, Carmencita; Hong Kong at night; and the Pacific Northwest–Lower Mainland, where he serves as Assistant Professor of geography at the University of British Columbia.

DAVID E. SOPHER usually hesitates when asked where he is from. His parents were of the Baghdadian Jewry; he was born and grew up in Shanghai; he thinks of Berkeley, where he earned his Ph.D., as a spiritual home; and he has subsequently resided for periods from two months to several years in Ahmedabad, Chicago, the Chittagong Hills (Bangladesh), Jerusalem, London, Mexico City, New York, Sacramento, St. Paul, San Jose, Sri Lanka, and Syracuse; the landscapes that appeal to him most, however, are Italy's. Home is presently "Upstate," where he is Professor of Geography at Syracuse University.

YI-FU TUAN, when asked which place he considers home, usually replies, modestly, "the earth." He has lived for a period of years in China, Australia, England, Canada, and the United States, and had shorter sojourns in the Philippines, Malta, Panama, Ivory Coast, and Buenos Aires. The natural landscape with which he feels the deepest kinship is the American Southwest. His favorite human landscape is now a wavering strip between downtown Minneapolis and the University of Minnesota where he is Professor of Geography.

Index